# BASIC PHYSICS

## Notes for Traffic Crash Investigators and Reconstructionists

## ABOUT THE AUTHOR

**R.W. (Bob) Rivers** is a graduate of Northwestern University Traffic Institute's traffic accident investigation and police management training programs. He completed training with the Canadian Institute of Science and Technology in technical mathematics and areas of physics, studied psychology at the Okanagan Regional College, completed police administration training programs through the Canadian Police College and the University of Minnesota, and patrol management with the IACP. He developed the traffic accident investigation and traffic law enforcement training programs of the Royal Canadian Mounted Police and course training standards for the Canadian Police College, University of Alberta, and the British Columbia Institute of Science and Technology in technical traffic accident investigation. During his 33 years service with the Royal Canadian Mounted Police, Inspector Rivers was employed extensively in general police work, highway patrol, accident investigation, research and planning, and training and development. Since his retirement, Inspector Rivers has authored various internationally-recognized textbooks, acted as a consultant and has assisted in traffic accident investigation training and research studies on an international basis (see web site: www.nanaimo.ark.com/rivers~/ ).

# BASIC PHYSICS

## Notes for Traffic Crash Investigators and Reconstructionists

An Introduction for Some
A Review for Others

*By*

R.W. RIVERS

CHARLES C THOMAS • PUBLISHER, LTD.
*Springfield • Illinois • U.S.A.*

*Published and Distributed Throughout the World by*

CHARLES C THOMAS • PUBLISHER, LTD.
2600 South First Street
Springfield, Illinois 62704



ISBN 0-398-07482-8 (spiral)

Library of Congress Catalog Card Number: 2003063359



*Printed in the United States of America*
*SM-R-3*

**Library of Congress Cataloging-in-Publication Data**

Rivers, R. W. (Robert W.)
Basic physics : notes for traffic crash investigators and reconstructionists : an introduction for some a review for others / by R. W. Rivers, Wade Bartlett.
p. cm.
Includes bibliographical references and index.
ISBN 0-398-07482-8 (spiral)
1. Traffic accident investigation--Mathematics. 2. Traffic accidents--Mathematical models. 3. Automobiles--Dynamics. I. Bartlett, Wade. II. Title.

HV8079.55.R54 2004
363.12'565--dc22

2003063359

# PREFACE

This manual is intended to provide a basic understanding of physics as applied to traffic crash investigation and reconstruction and to provide an understanding of the subject that will lead to advancement in this profession. It is intended to serve a need for investigators who possess a good knowledge and understanding of elementary algebra and trigonometry, and who have successfully completed at least an at-scene traffic crash investigation course and wish to further their knowledge towards competency in advanced traffic crash investigation and reconstruction.

The range of topics included are those considered to be fundamental and which best serve the purposes of illustrating the methods and procedures vital as an introduction to physics. Essentials of the subject as related to vehicle motion are stressed. The mathematics used is kept simple and in straightforward, easy-to-understand language. Comments and examples, and a very comprehensive list of terms and definitions, supported by many illustrations and diagrams, are provided to give the reader a unified view of basic physics, all areas of which can be easily located by using the detailed index. A goal for the reader in studying this material should be to learn to think analytically. Of utmost importance is not so much learning the answer to a particular problem involving physics, but rather a way of rationally working from knowns to determine the unknowns of any reconstruction project.

A study of this work will assist students and others who wish to increase their knowledge of physics in overcoming the drawbacks encountered with many otherwise excellently-prepared textbooks currently available which explain a given principle by an author who has an insight into the subject matter but which is not shared by the reader or student. In some of these instances, explanations are written in an abstract manner which leave the student confused as to the application of the principle. With the basic knowledge that can be attained with this manual, the reader will be in a position to both better understand and apply these principles.

Except for a few cases, derivations of the various formulae are not shown. However, in the case of each individual formula, all elements are explained in such a manner as to make known and easily understood all factors that are involved. Complete derivations for many of the formulae used in this manual can be found in the *Derivations Manual for Formulas Used in Traffic Accident Investigation and Reconstruction,* by Wiley L. Howell, and *Fundamentals of Applied Physics for Traffic Accident Investigators* (Vol. 1), authored by John Daily and Nathan S. Shigemura, both of which are published by the Institute of Police Technology and Management (IPTM), University of North Florida. (See the bibliography or http://www.iptm.org for details).

As a manual that is prepared for use on an international basis, all mathematical and/or technical references are shown in both the United States/English system and Metric (SI). To assist the reader and to avoid confusion, conversion factors and constants will be found throughout the book. In some of the calculations shown, however, exact conversions from the English system to SI are not made in order to avoid unnecessary use of multiple significant figures in the latter. The corresponding values that appear in these cases should therefore be considered approximations. To further assist the reader, examples are shown separately for each system and they should be treated as independent of each other.

When conducting some tests involving ratios, such as determining a coefficient of friction using a drag sled, many investigators wish to use a spring scale or similar device calibrated only in pounds or kilograms. In these cases, it should be understood that when reading this manual where it might be thought that in SI only newtons should be used, the important thing to do is to use the same type of units. *Example,* in the case of a drag sled, if the weight is measured in pounds (kilograms), the force (pull) to slide the sled along the surface should likewise be measured in pounds (kilograms).

R.W.R.

## ACKNOWLEDGMENTS

I wish to acknowledge my gratitude to the following professionals for their timely and helpful suggestions received during the various stages of manuscript development:

***Initial Draft Reviews***

Stephen R. Robinson, P.Eng., President, Robinson Forensic Engineering Inc., Moncton, N.B., Canada. Telephone: (506) 854-2526 Fax: (506) 852-2923. Cell: (506) 381-1804. E-mail: <RFEng@nbnet.nb.ca>.

Wade Bartlett, P.E., President, Mechanical Forensics Engineering Services, LLC, Rochester, NH. Telephone: (603) 332-3267. Fax: (603) 330-3347. E-mail: <wade@mfes.com> http://www.mfes.com.

John Ruller, Senior Traffic Accident Reconstructionist, Road Accident Investigation Service P/L, Bellbowrie, Qld., Australia. Telephone/Fax: 011-61-7-3202-7707. E-mail: <rjruller@eis.net.au>

***Final Draft Review***

John Duff, P.Eng., Ph.D., For-Eng Consultants Ltd., Sechelt, B.C., Canada. Telephone numbers: 1-800-461-8913 and 604-885-8963. Fax (604) 885-8967. E-mail: <for-eng@dccnet.com>.

R.W.R.

# THE METRIC (SI) SYSTEM)

The metric system, **Le Systéme International d´Unités** (International System of Units, abbreviated *SI* in all languages) is used in most countries outside the United States. Because this manual is prepared for international use, all mathematical formulae and problem-solving examples are shown in both the *United States/Imperial* or *English* and *Metric (SI)* systems.

In North America, a decimal fraction is generally indicated by means of a (decimal) point on the line (not a dot in the raised or centered position). In this regard, it is important for North Americans and many others to understand that in some countries it is the dot in the raised position that is used. Also, that in some countries a comma is used. It is the North American practice of using the dot as a decimal point situated on the line that is followed in this manual.

# DISCLAIMER

In the preparation of this manual, various published works and technical papers have been studied and consultations with experts in this field have taken place. The information and practices set out herein are, to best of the author's knowledge, experience, and belief, the most current and accurate in terms of providing a basic understanding of the physics used in traffic crash investigation and reconstruction. However, the author, publisher, and editors expressly disclaim all and any liability to any person, whether a purchaser of this publication or not, as a consequence of anything stated, done or omitted to be done, whether in whole or in part by such person in reliance upon any part of the contents of this publication. Every acceptable method, procedure or explanation may not be presented, and some individual cases or sets of circumstances may require additional or substitute procedures and explanations. Also, since statutes, ordinances, and organizational policies and procedures differ widely in various jurisdictions, those of the particular jurisdiction of the user should govern when there is any apparent conflict between them and the contents of this manual.

# CONTENTS

# BASIC PHYSICS

## Notes for Traffic Crash Investigators and Reconstructionists

# Chapter 1

# INTRODUCTION TO PHYSICS AND VEHICLE MOTION

## INTRODUCTION

**1.001** The following notes are intended to serve as an introduction for some and a review for others in relation to physics and its application to traffic crash investigation and reconstruction. The topics that are presented and the extent of their content are limited to that which is considered to be fundamental in terms of illustrating the methods and procedures vital as an introduction. The material presented should also pave the way for the study of more advanced works such as those found in the textbooks listed at the end of this manual.

## PHYSICS DEFINED

**1.002** The term *physics* is derived from a Greek word meaning nature. It explains and quantifies the physical world around us. Physics, as an exact science, deals with matter and energy in terms of motion, and is knowledge possessed as the result of study and practice, which is classified and accumulated.

## MECHANICS AND DYNAMICS

**1.003** The part of physics of special interest in traffic crash investigation and reconstruction is under what is called *mechanics.* We may define mechanics as (a) *that branch of physics dealing with the motions and states of material bodies,* or (b) *that area of physics that deals with motions and forces that move bodies through direct contact.* One exception to this, however, is the force of gravity which does not have to be in direct contact with a body to produce a motion.

Mechanics is usually divided into two parts: (1) *Dynamics* and (2) *Kinematics,* defined as follows:

1. *Dynamics.* The branch of physics that specifically deals with the influences of forces on motion and may be further divided into two parts: (a) *statics* and (b) *kinetics.*
   (i) *Statics.* The branch of physics dealing with conditions in which bodies are kept at rest by the actions of forces. The words *static* and *equilibrium* describe the forces acting in such a way that the net force on an object is zero. A body in equilibrium and at rest is said to be in static equilibrium, whereas a body moving with uniform velocity (constant speed) is said to be in dynamic equilibrium.
   (ii) *Kinetics.* The branch of physics dealing with the effect of forces in the production of modification of motion in bodies.
2. *Kinematics.* The branch of physics that deals with (a) *analytical and mathematical descriptions of all kinds of motion;* (b) t*he quantitative description of motion; involving only the description of motion and not concerned with the causes of motion.*[1] Kinematics is considered to be the starting point for mechanics.

## NEWTON'S THREE LAWS OF MOTION

**1.004** The principles contained in *Newton's Three Laws of Motion* form the basis of the whole subject of *mechanics* and should be thoroughly understood by the student and investigator. The three laws are:

Law 1: *Everything remains at rest or moves with constant velocity (in a straight line) unless acted upon by an external unbalanced (not counteracted) force.*

Or, stated another way:

> *A body at rest remains at rest, and a body in motion remains in motion with constant velocity, unless acted upon by some external force.*

This First Law is commonly called the *Law of Inertia.*

Law 2: *A free body acted upon by a constant force moves with constant acceleration in the direction of the force. The amount of acceleration experienced by a body is directly proportional to the acting force and inversely proportional to the mass of the body.*

Or, stated another way:

> *The time rate of change of the momentum of a body is proportional to the net external force acting upon the body and is in the direction of this force.*

Law 3: *To every action there is an equal and opposite reaction.*

Or, stated in other ways:

> *Whenever one body exerts force upon a second body, the second body exerts an equal and opposite force on the first.*

> *To every action there is an equal and contrary reaction.* (Here the term *action* is used to imply *force.*)

It should be noted that *action* combines force and time. The force exerted by one body on the other is equal and opposite to the force exerted by the other body on the first body. And, of course, when the first body is in contact with the second, the second must be in contact with the first. The *force x time* concept is important when dealing with collisions, and is known as *Impulse.*

## FORCE DEFINED

**1.005** *Force* is that which can impose a change of velocity on a material body, its direction of motion, or both as explained by Newton's Second Law of Motion, often referred to as the principle of inertia (meaning idleness or laziness). The principle of inertia requires that in the absence of a force, an object will remain at *rest* if at rest, or at a *constant* velocity if it begins in motion. We use the term *mass* to quantify a body's resistance to having its velocity, direction or magnitude changed. Force is expressed algebraically as:

***Formula 1–01A***

$$F = Ma$$

or

***Formula 1–01B***

$$a = \frac{F}{M}$$

where $F$ = the net external force acting on the body
$a$ = acceleration
$M$ = mass

Newton's Second Law says acceleration $a$ is directly proportional to the force $F$ acting on a body and is inversely proportional to the mass $M$ of the body. Expressed with the use of symbols, this statement would be written $a = F : M$, the colon standing for *is proportional to.*

A proportion of this kind is a ratio (comparison) of the number of units in one quantity to the number of units in another. The mathematical operation used to calculate a ratio is division. Therefore, the above statement can also be written as:

$$a = \frac{F}{M}$$

where the right side of the equation appears as a fraction and is easier to work with.

### *Derivation*

Start with

$$a = \frac{F}{M}$$

You are going to isolate the force **F**.

Multiply both sides by the mass **M**.

$$\frac{M}{1} x \frac{a}{1} = \frac{F}{M} x \frac{M}{1}$$

Cancel terms.

$$\frac{M}{1} x \frac{a}{1} = \frac{F}{\not{M}} x \frac{\not{M}}{1}$$

Rewrite.

$$\boldsymbol{Ma = F}$$

Rearrange.

$$F = Ma$$

Derivation source: Howell, Wiley.: *Derivations Manual.* Institute of Police Technology and Management, University of North Florida, Jacksonville, Florida, 1994, pp. 34, 35 (Reproduced with permission).

If all outside forces acting on a body cancel out each other when added together, the net force acting on the body is zero, and it does not accelerate, e.g., a vehicle at rest on a roadway (see Fig. 1–01)–since the vehicle is not accelerating up or down, we know vertical forces are balanced: $W = F_1 + F_2$. If all outside forces added together do not cancel out each other, they can be said to be unbalanced, and the net force causes a proportional acceleration.

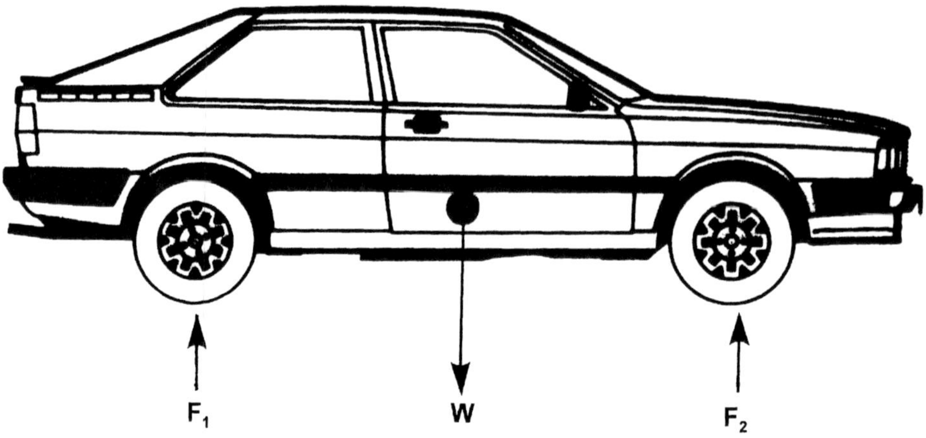


Figure 1–01.

*Note:* Force is discussed in much greater detail throughout this manual (see particularly Chapter 8).

## FIRST LAW

### *Inertia*

**1.006** All matter has mass and inertia. Inertia, from a Latin word meaning idleness or laziness, relates to the force necessary to change an object's motion and is defined as:

> *The property of matter which requires that a force be exerted on a body to accelerate it.*

It is the tendency for any object to resist change in velocity, magnitude, or direction. That is to say that if the object is at rest, it tends to stay at rest, and if the object is in motion, it tends to maintain the same speed and direction. If it is in motion, inertia is that property of matter by which it maintains a constant velocity in the absence of an unbalanced external force, and continues in motion in a straight line with constant velocity until some external force changes its state of motion. It is the mass of the object that determines the quantity of inertia that is possessed by a body.

### *External Forces*

**1.007** When two bodies collide, each body exerts an external force on the other, thereby causing a change of velocity in each body.

a. All objects, or bodies, consist of a collection of atoms. Every atom has a very small amount of mass. It is this collection of atoms that constitutes the mass of the object. The greater the number and size of the atoms, the greater is the mass. From Newton's Second Law, as the mass is increased, the greater will be the force necessary to generate a unit value of acceleration.

b. Inertia is a term that means that a body is reluctant to change its state, i.e., to change its velocity. It is identical to the mass of the body when dealing with translational (straight line as distinct from rotational) motion. Inertia and mass can be used interchangeably for translational motion. It has the units of pounds mass in the English system and kilograms in SI.

### *Curves*

**1.008** When a vehicle is in motion, there is the tendency for it to remain in motion and to continue in a straight line. In the case of traveling at an excessive speed in a curve, the result could be that the vehicle would sideslip and leave the highway tangent to the curve (in a straight line). In the case of a vehicle traveling in a curve, an occupant tends to be pushed towards the outer side of the vehicle, or perpendicular to the direction the vehicle would have traveled in a straight line if it had not followed the curve.

## SECOND LAW

**1.009** Newton's 2nd Law proves to be a powerful tool for understanding natural phenomena. For any body, it relates the external force acting on the body, its mass, and its acceleration. If any two are known, the third can be found; thus:[2]

a. If we know the resultant of all the forces acting on a body as well as the body's mass, we can deduce its acceleration and predict its motion.

b. If we know the forces acting on a body and can measure the body's acceleration, we can calculate its mass.

c. If we know the mass of a body and can find the body's acceleration by observing its motion, we can determine what net force is acting on the body. Even if we do not know the mass, from our knowledge of the motion we can infer the ratio of force to mass.

## THIRD LAW

**1.010** Examples of the Third Law are:

a. The wheels of a vehicle traveling along a roadway push backward on the roadway surface, but at the same time, the roadway surface pushes forward on the wheels with an equal force.

b. The weight of a vehicle causes the tires to push downward on the roadway surface with a force equal to that which the surface exerts upward on the tires.

c. When tires are skidding on a roadway surface, they exert a force on the roadway surface. At the same time, the roadway surface exerts a force on the tires that is equal in magnitude but opposite in direction.

d. A semitrailer pulls back on a truck tractor with exactly the same force as the truck tractor pulls forward on the semitrailer.

e. When two vehicles collide with each other, the force of vehicle 1 against vehicle 2 is equal and opposite to the force of vehicle 2 against vehicle 1, even if one is stationary and massive and the other is light and moving quickly.

f. When a vehicle collides with a utility pole, the forces sustained by both objects are equal and opposite, though the effects on each may be quite different.

# Chapter 2

# MEASUREMENTS IN PHYSICS

**2.001** Two systems of measurement are in common use around the world: (1) the British Engineering System, most often referred to as the English or Imperial system, and (2) the metric system, properly known as the International System of Units after the French name, Système International (SI). In the English system, the fundamental units are pounds for force, feet for length, and seconds for time. In SI, it is the kilogram (kg) for mass, meter for length, and seconds for time. Other units defined in terms of these base units are called derived units.[3]

**Table 2–01**

**Derived Units**

| *Quantity* | *English Symbol* | *SI Symbol* |
|---|---|---|
| **Speed** | | |
| Miles per hour | *mph* | |
| Kilometers per hour | | *km/h* |
| **Velocity** | | |
| Feet per second | *ft/s* | |
| Meters per second | | *m/s* |
| **Force** | | |
| Pound force or pound mass per second squared | *lbf or lbm* $ft/s^2$ | |
| Kilogram meter per second squared or Newton | | $kg\text{-}m/s^2$ *or N* |
| **Pressure** | | |
| Pounds per square inch | $lbf/in^2$ *or psi* | |
| Newton per square meter or Pascal | | $N/m^2$ *or Pa* |

**2.002** Historically, the SI system used two systems of basic units. These were the MKS (meter, kilogram, second) system and CGS (centimeter, gram, second) system. Some scientists still refer to SI units as MKS units.[4]

## ACCELERATION RATE

**2.003** *Acceleration* is defined as *the time rate of change of velocity.* In a scientific sense, acceleration can be said to be either positive acceleration (+) or negative acceleration (–). Contemporary usage of the terms in traffic crash investigation is simply *acceleration* for positive acceleration and *deceleration* for negative acceleration. These two terms are easily understood by everyone with no additional explanation. When the initial or final velocity is zero *(0),* the average acceleration rate can be shown algebraically as follows:

***Formula 2–01 (Initial or Final Velocity = Zero)***

$$a = \frac{V}{t}$$

where $a$ = constant or average acceleration rate
$V$ = velocity
$t$ = time in seconds

**2.004** When an object is traveling at a known velocity and then for a period of time its velocity changes (±), such as when a vehicle accelerates to pass another vehicle, or decelerates for safety reasons, the following formulae can be used to calculate its rate of acceleration or deceleration:

***Formula 2–02A (Acceleration Rate)***

$$a = \frac{V_f - V_o}{t}$$

where $a$ = acceleration rate
$V_f$ = final velocity

$V_o$ = original velocity
$t$ = time in seconds

*Note:* This formula is valid only for a constant rate of acceleration. Continued acceleration of a vehicle results in a lowering of the rate of acceleration, say, after a few seconds. As engine rpm approaches the red line, torque falls off, resulting in a lowering of the rate of acceleration. Any gear changes will also lower the acceleration rate because of the time required to make the change.

***Formula 2-02B (Deceleration Rate)***

$$a = \frac{V_o - V_f}{t}$$

where $a$ = acceleration/deceleration rate
$V_o$ = original velocity
$V_f$ = final velocity
$t$ = time in seconds

## THE POUND

**2.005** By an act of Congress of the United States, the *standard pound* is defined as having 0.45359237 times the mass of the standard kilogram. The *pound,* the basic unit of force in the English gravitational system, is defined as: *The weight of the standard pound at sea level and 45° latitude.* The conditions of sea level and 45° latitude are necessary because the strength of gravitational attraction varies with changes in altitude and latitude. However, for purposes of traffic crash investigation, we may consider the gravitational pull of the Earth to be a constant, and while the distinction between mass and weight may thus seem academic, the distinction is vital when one considers the importance of *force* as a factor in these investigations.

## THE SLUG

**2.006** The slug (sluggish = inertial), is defined as follows:

> *One slug is the mass of a body which experiences an acceleration of one foot per second per second when acted upon by an unbalanced external force of one pound.*

Recall the formula $F = Ma$ (Formula 1.01A). A force of 1 lb gives a 1-lb mass an acceleration of 32.17398 ft/s$^2$ and gives a 1-slug mass an acceleration of 1 ft/s$^2$. The slug, therefore, has a mass 32.17398 times that of the standard pound mass. As is done in this manual, those who use the English system of units often replace $M$ in all formulas by $W/g$. In this way, the mass unit, i.e. slug, is avoided.

## NEWTON

**2.007** The newton is defined as follows:

> *One newton is that force which produces an acceleration of one meter per second per second in a mass of one kilogram.*

It is derived from the fundamental units of kilograms, meters and seconds. The newton is a so-called absolute unit of force because by definition, it already accounts for the pull of the earth. Using SI units in the same formula, $F = Ma$ (Formula 3-01), we find 1N = 1 kg-m/s$^2$. This is to say that the force of one newton, when applied to a mass of 1 kg, gives that mass an acceleration of 1 m/s$^2$. Incidentally, a 1 kg mass at sea level at latitude 45° would experience a force of 9.80665 N, because of the pull of the earth's gravity at that location. Thus, a scale that was calibrated to read in kilograms could also be calibrated to read in Newtons where 1 kg = 9.80665 N.

## VECTORS AND SCALARS

**2.008** In mathematical physics, we are concerned with two classes of quantities, namely: (1) vectors and (2) scalars.

1. A *vector* is usually represented by a directed straight-line segment whose length indicates the magnitude of its quantity. A vector must also have direction to describe it completely, usually denoted by an arrow pointing in the direction in which the vector acts. Vector quantities as a group include displacement, velocity, force, acceleration, momentum, force moment, and angular velocity. Vector quantities are not subject to the rules of algebraic analysis but may be manipulated under a system called vector analysis (see Fig. 2–01).

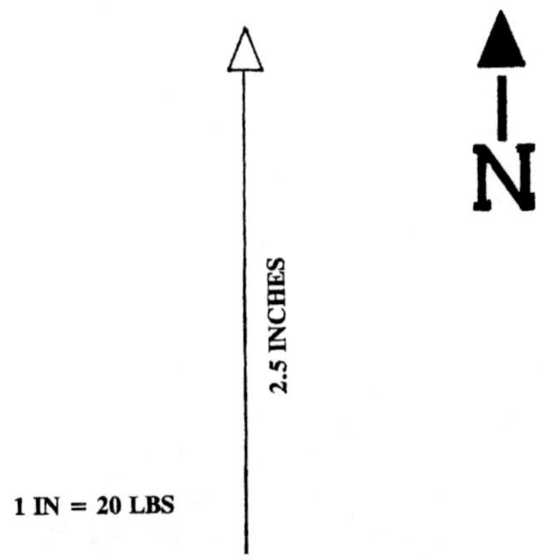


Figure 2–01. Arrow (left) represents a 50 lb force acting in a Northward direction. Length of arrow, drawn to scale, represents the magnitude of the vector quantity (50 lb). At right is the north arrow. Scale = 20 lb/inch force (or equivalent SI).

2. A *scalar quantity* has no direction and can be specified by a single number, the magnitude, with appropriate units. Scalar quantities include length, area, volume, capacity, density, energy, mass, and time.

**2.009** *Resultant.* Two or more forces may act at one time upon an object. These forces are the component vectors, or components, and their combined effect is the resultant vector, or simply the *resultant* (see Figs. 2–02 & 2–03). It is possible to have a single vector represent the addition and/or subtraction of a number of individual, similar vectors, e.g., force vectors. The *resultant* is that single vector which would have the same effect as all the original vectors taken together. For example, when two or more forces act concurrently (simultaneously) at a point, the *resultant force* is the same as a single force being applied at the same point producing the same effect.

### *Example*

Two men attempt to move a car by pulling on a rope that is attached to the vehicle's bumper. They both pull in the same direction. One man exerts a force (pull) of 60 pounds, the other a force of 80 pounds. We can diagram the forces involved using a scale of 1 in = 20 lb. Since the forces are in the same direction, we draw the arrows head-to-tail. The resultant force being exerted is represented by the combined length of the arrows, which is seven inches or 140 lb of force (see Fig. 2–02 ).

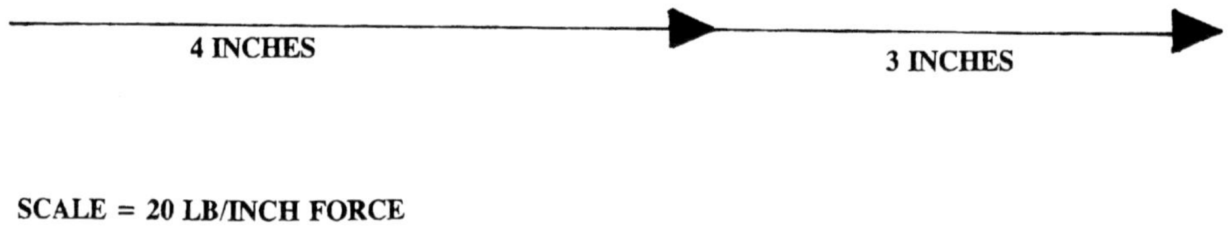


Figure 2–02. Vectors acting on the same point in the same direction are summed directly (arithmetically) to find the resultant. In this example, the resultant is a 140 lb force (or equivalent SI).

If the circumstances were such that the men were pulling at an angle to each other, one pulling north with a force of 80 lb and the other pulling east with a force of 60 lb, we would then have multiple vectors acting at an angle, so that the vectors could not simply be added together to obtain the resultant. We could, however, draw a vector diagram to find it. Again using the scale of 1 in = 20 lb of force, we would have a 4-inch arrow pointing north and a 3-inch arrow pointing east. In this case, the resultant can be determined geometrically by constructing a rectangle whose sides equal the magnitude of the component vectors, i.e., 80 lb (4 in) and 60 lb (3 in). The diagonal will then equal the sum of the two forces, or simply the resultant. We find in this case that the diagonal is 5 inches long. The resultant is therefore (5 lb x 20 lbs) 100 lb (or equivalent SI). A protractor can be used to determine the specific direction of the resultant. In this instance, it is 37° east of north (see Fig. 2–03).

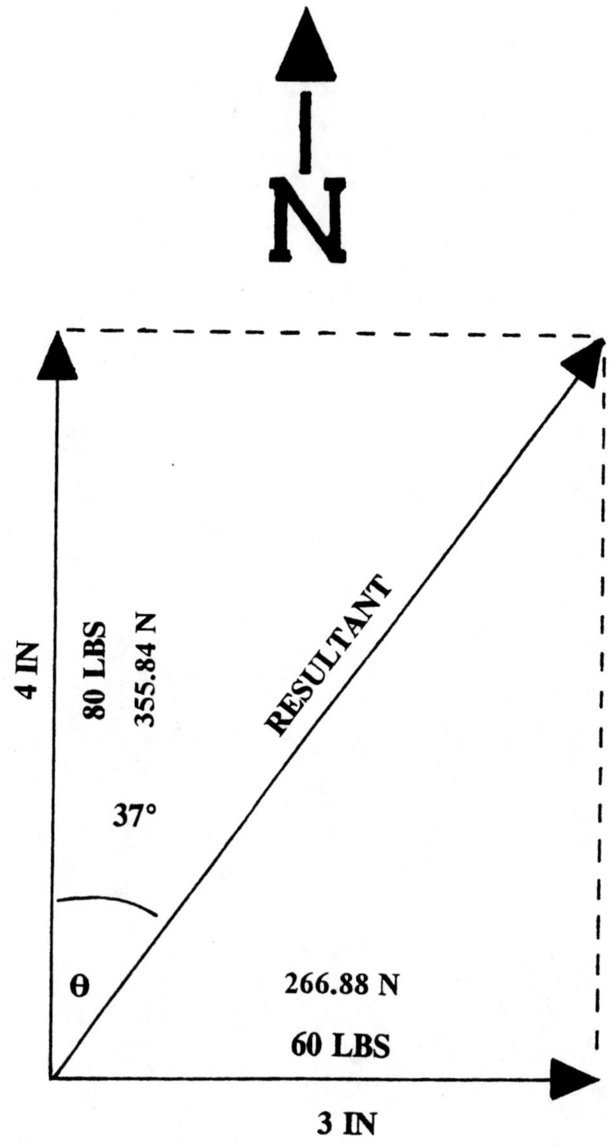


Figure 2–03. When vectors act at an angle to each other, their net effect or resultant is their geometric sum.

# Chapter 3

# GRAVITY, MASS AND WEIGHT

## GRAVITY

**3.001** Gravity is a force defined in Newton's Law of Gravitation:

> *Any two bodies in the universe attract each other with a force that is directly proportional to their masses, and inversely proportional to the square of their distance apart.*

For our purposes, gravity is the constant vertical attraction (force) of an object downwards towards the center of the earth, derived from one pound of force acting for one second upon a mass of one pound or one kilogram, imparting to this mass a velocity of 32.174 feet per second or 9.80665 meters per second, at sea level, and described by the acceleration it causes the object. The value conventionally taken for standard acceleration of gravity, namely, 32.174 $ft/s^2$, is derived by rounding to the nearest 0.001 $ft/s^2$ after converting from the internationally accepted metric value of 9.806 65 $m/s^2$. The force of gravity is expressed in pounds in the English system (used in the United States) and newtons in the metric system (SI).

**3.002** In terms of acceleration due to gravity, all objects fall at the same rate if air friction is neglected. The *constant rate* nomenclature refers to the fact that the rate does not vary during the fall of an object rather than to the common rate for all objects. This commonality of rate is caused by the fact that the more massive the object, the larger will be the attractive force. However, from Newton's Second Law, acceleration is inversely proportional to mass, so the more massive the object, the slower the acceleration for the same attractive force. The two effects cancel each other out, resulting in all objects falling at the same rate. The velocity at which objects fall increases due to the constant acceleration. At the end of the 1st second, an object will have fallen 16.1 ft (4.905 m). Its velocity will be 32 ft/s (9.81 m/s). At the end of the 2nd second, it will have fallen $2^2$ or 4 times further, i.e., 64.4 ft (19.62 m), and it will be traveling twice as fast, i.e., 64.4 ft/s (19.62 m/s). At the end of the 3rd second, the distance will be, $3^2$ or 9 times the 1st second distance and 3 times the 1st second speed and so on. An example of these computations follows (see also Fig. 3–01).

*Example–Distance traveled in 1st second*

**Formula 3–01**

$$d = \frac{at^2}{2}$$

where $d$ = distance in ft (m)
$a$ = rate of acceleration due to gravity
$t$ = time in seconds

| *English* | *SI* |
|---|---|
| $d = \frac{at^2}{2}$ | $d = \frac{at^2}{2}$ |
| $d = \frac{32.2 \times 1^2}{2}$ | $d = \frac{9.81 \times 1^2}{2}$ |
| $d = 16.1\ ft$ | $d = 4.91\ m$ |

**3.003** When an object falls through air, the actual rate of fall depends upon the object's mass and shape because of the effect of air drag. As the object falls, it must push the air aside. Compare the fall of a distinctly light body, such as a soap bubble, with the fall of a solid object, such as a billiard ball, where both are of similar size but have considerably different masses. If both were dropped at the same time from the same height, the bubble would, of course, not fall as fast as the billiard ball. The only time the rates of fall of these two different types of objects would be the same would be in a perfect vacuum, where there is no air resistance.

Another point to be considered, however, is that gravity acts independently of outside forces. In Figure 3–01, for example, if Vehicle 1 left the highway at a speed of 80 mph (129 km/h) and was catapulted horizontally over a 100 ft (30 m) cliff at the same time Vehicle 2 was pushed over the edge of the cliff, they would both strike level ground at the same time. This would occur because as Vehicle 1 leaves the edge of the cliff, it would have the same vertical (downward) acceleration as Vehicle 2, namely, 32.2 ft/s$^2$ (9.81 m/s$^2$), caused by the gravitational pull of the earth.[5]

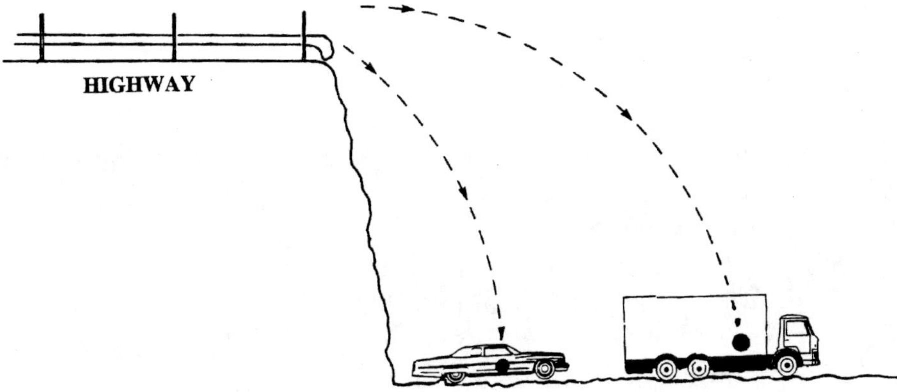


Figure 3–01. Acceleration of vehicles during a fall.

**3.004** It is customary in most contemporary works to round up the acceleration-due-to-gravity constants *32.174* and *9.806 65* to **32.2** and **9.81**, respectively. These latter constants are considered acceptable for all points on Earth.

**3.005** One *g*-force is equal to a mass accelerated at the acceleration rate of gravity, i.e., 32.2 ft/s$^2$ (9.81 m/s$^2$). The gravitational force *(F)* that the Earth exerts upon any body on its surface is equal to *Mg*, expressed as:

***Formula 3–02***

$$F = Mg$$

where $F$ = force
$M$ = mass
$g$ = acceleration due to gravity

For a freely falling body, the downward force *(F)* is equal to the weight of the body and the downward acceleration is equal to *g* or:

***Formula 3–03***

$$F = Ma$$

where $F$ = force
$M$ = mass
$a$ = acceleration due to gravity

**3.006** The value of gravitational pull *g*, depends upon the distance the body is from the center of the Earth. The value decreases as the distance increases. For example, at the equator, being the greatest distance on Earth from the Earth's center and where the latitude is 0°, the *g*-force is 9.78039 $m/s^2$. As the latitude increases, the acceleration of gravity increases until at 90° (the pole) the *g*-force is 9.83217, a change of 0.0517 $m/s^2$ and so insignificant that it can be disregarded for purposes of traffic crash investigation. If two vehicles having the same mass were located together at a point approximately 1,600 miles (2,574 km) above the earth, their combined weight, because of the lesser influence of Earth's gravitational pull, would be only about that of a single vehicle setting on the Earth's surface. However, when it comes to two vehicles of equal mass, one of which is traveling along a highway at sea level and the other traveling over a high mountain pass, taking the difference in their distances from the Earth's center into account would have a negligible effect on any traffic crash calculation.

**3.007** Aside from gravitational force, which directs an object downward, any other force that is placed on an object may have an effect of its own on the object. A body propelled upward, as in a flip and vault crash case, moves in a direction opposite to the gravitation force present, but will eventually stop and be pulled back toward the ground by the force of gravity–unless, of course, its velocity is sufficient to cause it to defy the earth's gravitation pull and go into orbit around the Earth. Understandably, such escape velocity cannot be reached by motor vehicles.

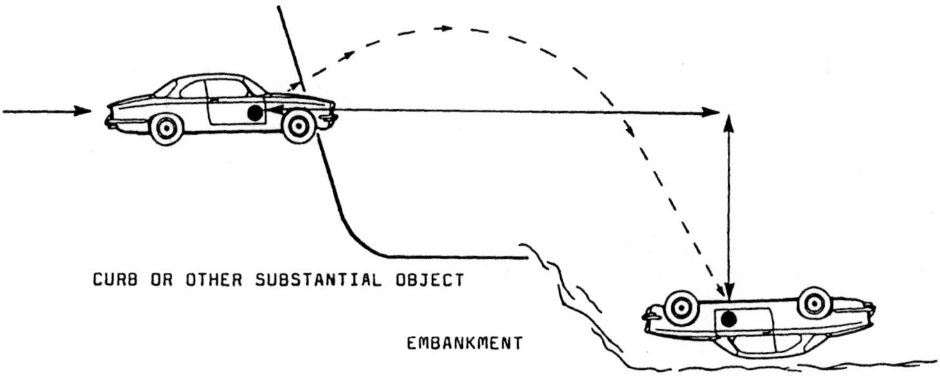


Figure 3–02. A flip and vault with the landing lower than takeoff.

## MASS

**3.008** Mass is the amount of matter in a body and is constant anywhere in the known universe; at least it does not vary appreciably until it approaches the speed of light [186,262 miles per second (300,000 km/s)]. The difference in mass between substances having the same volume relates to the composition of their molecules and atoms. Heavy elements, such as plutonium, which is used in atomic reactors, have large numbers of electrons closely packed together. Light elements, such as the gas hydrogen, have smaller numbers of atomic particles widely dispersed. Mass has two aspects, (1) inertia and (2) gravitation, and is commonly measured by weighing, i.e., by comparing the gravitational attraction of the body whose weight is to be determined to the Earth. If an object weighs one pound at the Earth's surface, what is meant is that this is the amount of mutual attraction between the object being measured and the Earth at that location. When measuring mass, the unit of mass used must be such that it will give the answer in the correct units when used in the formula for Newton's Second Law of Motion. On Earth, the mass *(M)* of an object is its weight (W) divided by the gravitational pull or acceleration (a) of the Earth, i.e., gravity:

***Formula 3–04***

$$M = \frac{W}{g}$$

where $M$ = Mass
$W$ = weight
$g$ = acceleration due to gravity

Cross multiplying, we have weight = mass x gravity ( acceleration due to gravity), or $W = Mg$.

## WEIGHT AND MASS

**3.009** Considerable confusion exists in the use of the term weight to mean either the force of gravity or the force of mass. The investigator should be aware of the difference between mass and weight, even though the two may be treated as being synonymous without difficulty. Their use in a problem or example, however, must at all times be properly defined or understood. Bueche (1979) states that mass is a measure of the inertia of an object, while weight is the force of gravity on the object. They are not the same in either concept or units. However, they are related. If an object of mass $M$ falls freely toward the earth under the pull of gravity (a force $W$, the object's weight), then the acceleration of the object is the free-fall acceleration. For a freely falling body, $F = Ma$ becomes $W = Mg$,[6] the latter expressing the relationship that exists between the weight and mass of an object.

**3.010** Weight is the force with which an object is attracted to the Earth by the gravitational pull of the Earth. It is a force like any other force, vertical and unavoidable, and has the units of a force. Newton's Second Law indicates that weight *(W)*, being a force, must be a mass multiplied by an acceleration (see Para. 3.008). The acceleration is, of course, that which is produced by the Earth's gravitational field. When we say, for example, that a vehicle weighs 2000 lb or equivalent newtons *(N),* we mean that its mass is attracted by the Earth, such that a force of 2000 lb *(N)* is required to support it stationary at the Earth's surface. It should be noted here that the term *weight* can be ambiguous, depending on the context. The *weight* of an object depends on what is meant by *weight.* In the English system, it is usual and acceptable to refer to it in pounds. In metric, however, the general public would say that the object weighs, for example, 80 kg–here, *weight* means *mass.* The physicist may, however, say that the object weighs about 784 N–here, *weight* would mean the force of gravity. It is therefore suggested that the word *weight* be used only in situations where its meaning is perfectly clear within the context in which it is used, a policy that is followed throughout this manual.

**3.011** To further explain the distinction between mass and weight, consider two identical vehicles that together have twice the mass of one of the vehicles by itself. Their combined mass would remain the same regardless of the location of the vehicles in the universe. Their combined weight, however, could vary. For example, if one of the vehicles was moved further away from the Earth, *(g)* the force of the Earth's gravity on that vehicle would go down, the less that vehicle would weigh, and consequently the less would be the combined weight of the two vehicles. For our purposes here on the Earth's surface, we may nevertheless consider the gravitational pull of the Earth to be a constant, and while the distinction between mass and weight may thus seem academic, the distinction is vital when one considers another important factor in motor vehicle collisions–force.[7]

**3.012** It is commonly heard that a truck is twice as massive as a car. This is a correct statement inasmuch as when dealing with collisions, it is the mass of the vehicle that counts–not the weight. However, most often the intent is to say that a truck is twice as heavy as a car, particularly when the weights in question are a licensed gross weight as set out in a vehicle registration and a weight of a commercial vehicle and load as determined by a scale. Again, caution should be used to ensure that when giving testimony or preparing a report, no ambiguity exists when using the terms *mass* and *weight* (see Para. 3.010).

**3.013** When measuring mass, select the unit of mass that will give the answer in the appropriate units when used in a formula based on Newton's Second Law of Motion:

***Formula 3–05***

$$A.\quad M = \frac{F}{a}$$

or

$$B.\quad M = \frac{W}{g}$$

where $M$ = mass in pounds or kilograms
$F$ = normal force in pounds or newtons
$a$ = acceleration in ft/s$^2$ (m/s$^2$)
$W$ = weight
$g$ = acceleration due to gravity

**3.014** Formulae 3–05 and 3–06 can be used to show the relationship between mass and weight. It should be noted, however, that in SI, weight shown in kg is in reality the mass of the vehicle; therefore, in the following SI examples, there is no need to divide or multiply (as the case may be) the weight by the acceleration due to gravity. Further, it should be noted that in those countries where the use of SI units is relatively new, an object's weight is often expressed in kg. For example, apples may be sold by the kilogram and thought of as the weight of the apples. In fact, the purchase is apples having a mass of 1 kilogram. Another example is when a vehicle's weight is described as 1250 kg it is in fact the vehicle's mass that is being described. In Australia and many other countries, there is no room for confusion inasmuch as a vehicle's specifications will read, for example, 1250 kg mass.

***Example 1***

A vehicle weighs 2,000 lb (907 kg). Its mass would be:

*Formula 3–04*

$$M = \frac{W}{g}$$

*English* *SI*

$$M = \frac{2000}{32.2} \qquad M = 907\ kg$$

$$M = 62.11\ slugs\ (lb\ mass)$$

***Example 2***

A vehicle has a mass of 62.11 slugs (907 kg). Its weight would be:

***Formula 3–06***

$$W = Mg$$

where $W$= weight in pounds (newtons)
$M$= mass in slugs (kilograms)
$g$= acceleration due to gravity

*English* *SI*

$$W = 62.11\ x\ 32.2 \qquad W = (907\ x\ 9.81)\ 8897.67\ N$$

$$W = 2000\ lb$$

(For an explanation of this SI calculation, see Para. 3.010)

*Note:* Because of the confusion that sometimes exists between the terms *mass* and *weight,* one solution is to refer to *mass* as a measure of the amount of material in the object upon which no force is acting, when mass is intended, and to *force of gravity* as the *weight* when gravitational pull is intended. A force of gravity, pulling downwards, acts on each mass on the earth. Thus mass and force of gravity, while entirely different characteristics, are linked. Perhaps unfortunately, in everyday usage, weight is ambiguous in that it sometimes means mass and sometimes force of gravity, depending on the context, or the country in which one lives. It is very important for the investigator and the trainer in physics therefore to understand the difference between the uses of kg and mass when using formulae that require weight to be divided by gravity, in particular energy equations. If SI units are used, the information available may already, in fact, be described in mass units.

## CENTER OF MASS AND CENTER OF GRAVITY

**3.015** The generally accepted definitions for the center of mass and the center of gravity are:

a. *The point where a vehicle or other object would be in equilibrium or perfectly balanced if suspended by or at that point.*

b. *The point where the mass and the pull of gravity can be treated as being concentrated.*

In this manual, center of gravity *(CG)* and center of mass *(CM)* are considered identical and will be treated as one and the same unless the example or problem requires otherwise, at which time the difference will be indicated.

**3.016** The center of mass and the center of gravity are at the geometric center of a uniform, regular object, such as a sphere or brick, provided the object is either uniform in density or has density that changes in symmetrical fashion. In a billiard ball, for example, that point is precisely in the center of the sphere. In a hollow ball, it is also precisely in the center, but actually located in air. In the case of a vehicle, if the entire vehicle were in a uniform gravitational field, the location of the center of gravity would be identical to the location of the center of mass. However, the lower portion of a vehicle is a mite closer to the center of the earth than its upper portion, so the lower portion is more strongly under gravitational influence. The center of gravity is consequently located very slightly below the center of mass. While there is a difference, under ordinary conditions and for the purposes of traffic crash investigation and reconstruction, the difference in location is so ridiculously small that the two terms can be considered synonymous.

**3.017** The center of mass *(CM)* is in the location where any single external force that might be directed at or through that point, regardless of the angle (including an acute angle) from which it comes, will cause motion in the direction of the force but will not cause the object to rotate (see Fig. 3–03A). But if the force does not quite act in a direct line through the center of mass (an eccentric force), some kind of rotation of the object will be caused as during off-center side and front and rear-end collisions. For example, if an impact occurs on the left-hand side of a vehicle at a position forward of the *CM*, the vehicle will rotate clockwise (see Fig. 3–03B). Although the vehicle will rotate in these sit-

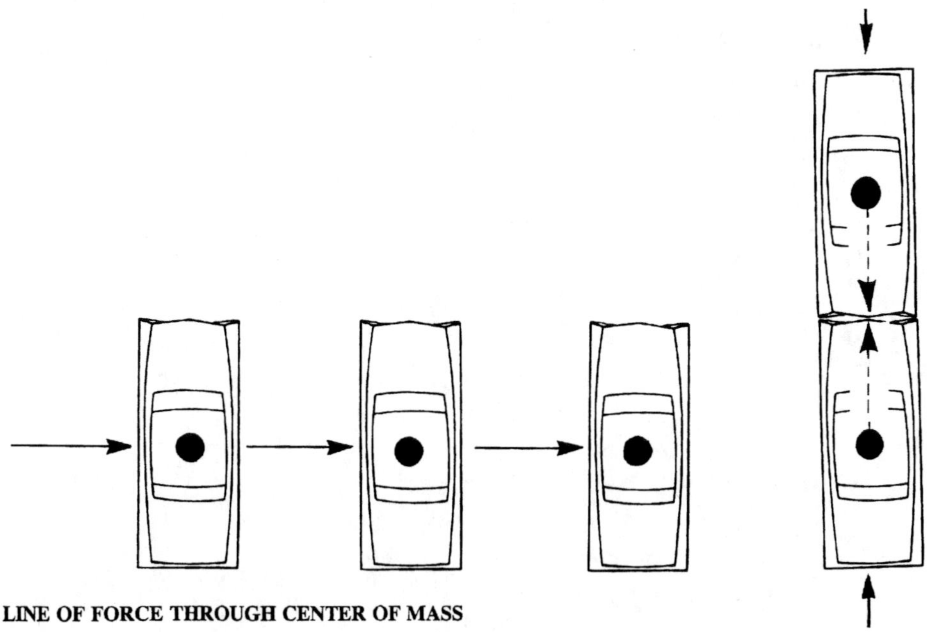


Figure 3–03A. Line of force through a vehicle's center of mass causes a translational movement, i.e., the vehicle does not rotate but rather moves along line of force.

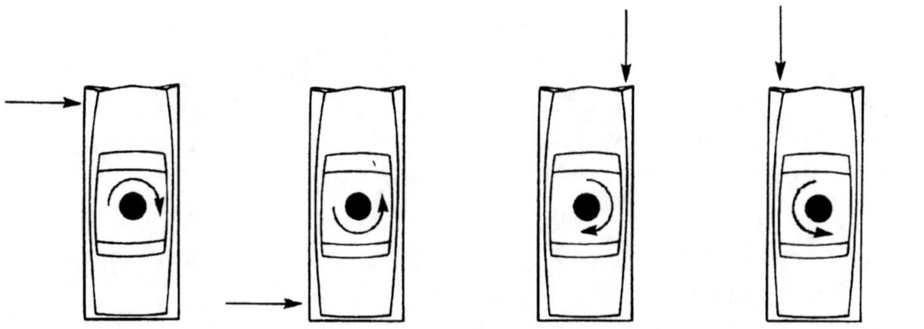


Figure 3–03B. Vehicles acted upon by an external, off-center (eccentric) force with subsequent rotation.

uations, the *CM* essentially continues in a straight path. The results of in-line and eccentric impacts could also be illustrated through the action of a cue on a billiard ball.

**3.018** In the case of a vehicle or other object that is projected or that falls, such as over a highway embankment, it is accelerated or pulled downward by the force of gravity. This force acts on the vehicle as if it were concentrated at a single point within the vehicle–that being the center of gravity (see also Para. 3.016).

**3.019** If a vehicle leaves the highway, such as when being catapulted or falling over a cliff or embankment, it continues to travel in the direction in which it left the highway. At the same time, however, it starts to drop (see Fig. 3–01). Its downward velocity increases with constant acceleration due to gravity, which acts as if it were concentrated at the point known as the center of gravity. However, its forward velocity stays essentially constant, i.e., it does not change, except for the effect of wind resistance.

## TIPPING

**3.020** When a vehicle is resting on its wheels in a stable position (its original base) and is struck from the side by another vehicle, with the principal direction of force being at or above the height of the center of mass of the (target) vehicle, the wheels on the side impacted may lift causing the vehicle to tip slightly. It may then return to its original stable position. With a sufficiently greater impact, the wheels on the side struck will lift higher, causing the vehicle to tip to a point of instability where it will overturn (see Fig. 3–04). This is because gravitational forces, although affecting all particles of a body, act as if concentrated at the center of gravity. When a vehicle is traveling in a curve or is struck on its side, and the tipping force is of sufficient strength to cause the center of mass to move to some point so that a plumb line suspended from it would fall outside the vehicle's original base, the vehicle will fall onto its side or overturn. If, however, the tipping force is not of sufficient magnitude to cause the center of mass to move so that a plumb line suspended from it would fall outside the original base, gravitational force will cause the vehicle to return to its former upright position.

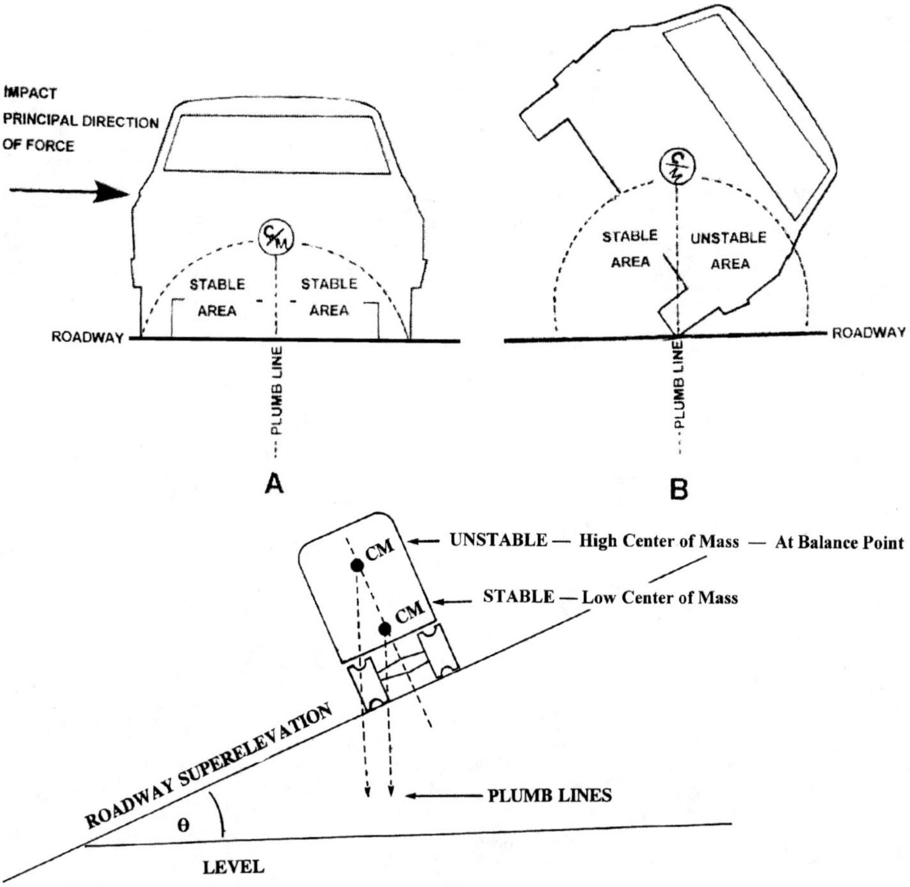


Figure 3–04. **A** represents an upright car, sitting on a level roadway in a stable position that experiences a side impact, the principal direction of force of which is slightly above its center of mass. **B** represents a tipped car at the balance point. If the car moves at all to either side, it will keep going in that same direction. **C** represents similar actions relative to a typical commercial-type vehicle.

## ACCELERATION, MASS AND FORCE

**3.021** Mathematically, Newton's Second Law can be expressed as:

***Formula 3–09***

$$a = \frac{kF}{M}$$

where $a =$ acceleration of a mass
$k =$ proportionality constant
$F =$ resultant of all forces which act upon the mass
$M =$ mass

The proportionality constant $k$ depends upon the units in which force, mass and acceleration are measured. By a simple choice of units, the constant k can be made equal to unity and become a constant of 1. When force, mass, and acceleration are measured in units for which k = 1, the Second Law can be rewritten as:

***Formula 3–10***

$$a = \frac{F}{M}$$

and,

***Formula 3–11***

$$F = Ma$$

where $F =$ force
$M =$ mass
$a =$ acceleration

From this it can be seen that:

a. If the value of mass *(M)* remains the same but the value of force *(F)* is increased, the value of acceleration *(a)* will also increase.

b. If the value of $M$ is increased, and the value of $F$ remains the same, the value of $a$ will decrease.

These are very important principles in the determination of speed based on the length of skid mark formulae.

The above equations tell us that the acceleration of a mass is proportional to the resultant *(F)* of all forces which act upon this mass. As an example, when a constant force *(F)* is applied to a vehicle of known mass, and that vehicle is on a level and frictionless roadway along which it can be accelerated in a straight line, a certain measurable acceleration is produced. If twice that force is applied and the acceleration is then measured, the acceleration will be found to be exactly double the first acceleration, and so on. When a known force produces a certain acceleration on a vehicle having a mass of 1,000 lb or kg., for example, and then the same amount of force is used on a larger vehicle having a mass of say 2,000 lb or kg, the acceleration will be only half as great as it was on the first vehicle. If the mass is increased to 3,000 lb or kg, the acceleration will be only one-third as great of that experienced by the first vehicle. If the experiment is continued with vehicles of various masses, it will be found that (within experimental error) the acceleration of the vehicles (or any other mass) will be directly proportional to the resultant force *(F)* and its direction will be that of the resultant force.

# Chapter 4

# MOTION

**4.001** Newton's 1st Law of Motion states: *Everything remains at rest or moves with constant velocity (in a straight line) unless acted upon by an external force.* Motion can therefore be defined as being a change of place or position. It is also a relative term. For example, a load upon or in a vehicle is in motion in relation to roadside objects as the vehicle travels along the highway, but it is at rest in relation to other things inside or upon the vehicle. (See also Frames of Reference, Para. 8.021.) Related terms and definitions include:

a. *Rectilinear motion.* Motion in a straight line.
b. *Curvilinear motion.* Motion along a curved path.
c. *Uniform motion.* Speed or velocity that is constant, i.e., uniform.
d. *Variable motion.* When the distances traveled in each succeeding unit of time are unequal.

In most motions that we observe, the velocity is not constant. In cases where its direction is variable, it is known as *accelerated motion.* Another kind of motion is that which is known as *constant acceleration* which can hold for a limited time, such as when a motor vehicle accelerates from a stop at a traffic signal and eventually reaches its final speed. When a vehicle is driven in city traffic, its velocity is continually changing–sometimes in magnitude, sometimes in direction, and sometimes in both, such as when a vehicle accelerates around a curve.

*Note:* Before proceeding further, the difference between the terms velocity and speed under this heading, and as they are commonly used in accident reconstruction, must be clearly understood, i.e., *velocity is specified by both a magnitude and a direction whereas speed is specified by only magnitude.* Newton's 1st Law of Motion states: *Everything remains at rest or moves with constant velocity (in a straight line) unless acted upon by an external force.* It is force that sets in motion a stopped or parked vehicle, and that accelerates or decelerates a moving vehicle. This force can be as the result of the power generated by the vehicle's motor or by a collision with another vehicle or other object; or as the direct result of the force of gravity, such as when an un-braked, parked vehicle on an incline begins to roll when no other force is present.

## UNIFORM CIRCULAR MOTION

**4.002** When a vehicle moves in a circular path at a constant speed, the velocity is constantly changing since the direction of the vehicle is constantly changing. Changing the velocity requires acceleration. Related terms and definitions include:

> *Centripetal force.* (a) A force attracting a body toward a center around which it revolves; (b) A force that is required to hold a moving object in a circular path, such as a curve in a highway.
>
> *Centrifugal force.* (a) The inertial reaction by which a body tends to move away from the center of rotation; (b) force that is the reaction to, or equal and opposite to centripetal; (c) an object's inertial resistance to being caused to move in a curve path.

*Note:* In a truly scientific sense, a *centrifugal force* is a fictitious force meant to show that which has the opposite effect of centripetal force–in reality, *inertia*. In a contemporary sense, it is an acceptable, relative term that is recognized and understood internationally for its intended purpose, and one that is widely used in a variety of issues relating to traffic crash investigation and reconstruction. (Centripetal and centrifugal forces are further explained in paragraphs 8.017–8.022.)

**4.003** In terms of circular motion, a mass undergoing a change in velocity has a force acting upon it. It follows therefore that an external unbalanced force is acting on a body to change it to move in a circular path. When this principle is applied to a motor vehicle traveling in a curve, the force of friction between the tires and the road keeps the vehicle moving in a circular path around the curve. If the speed is too great, the vehicle will have a tendency to sideslip, leave critical speed scuff marks, and depart the roadway tangent to the curve (see Figs. 4–01 & 4–02).

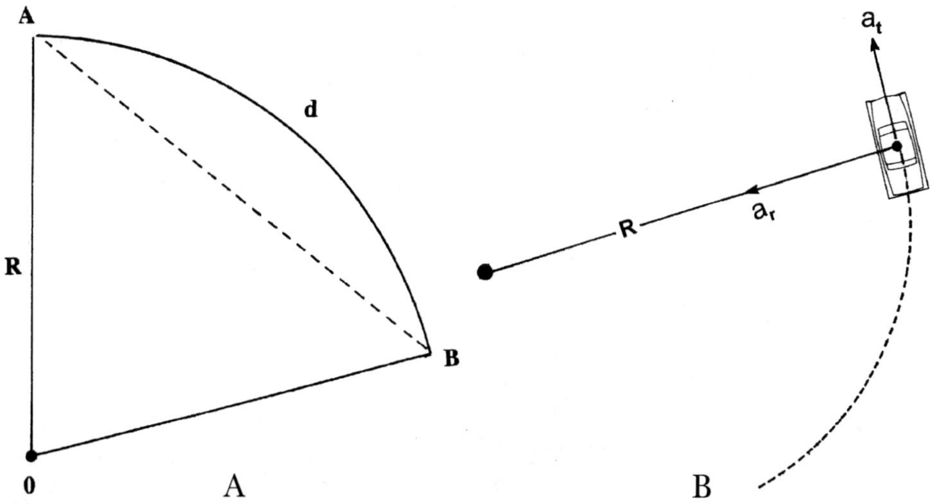


Figures 4–01A & B. Diagrams depicting a vehicle traveling in a circular path (see Para. 4.004).

**4.004** In the arc depicted in Figure 4–01A, *R* is the radius of the path followed by a vehicle in a curve. If we assume the time interval to be quite short, the length of the arc of the circle from *A* to *B* will be almost the same as the length of the chord from *A* to *B*. By using Formula 4–01, we can relate the acceleration of a body to the radius of the path and the squared velocity of the body, which is called *centripetal acceleration,* and which is directed toward the center of the circular path followed by the unit (see Fig. 4.01B, Formula 4–01 & Example 1). It should be pointed out that Formula 4–01 deals with a constant speed and a constant radius. At a velocity where the vehicle can no longer be steered to a tighter path, and since deceleration forces are small and there are no other forces acting to change the direction of the vehicle, the vehicle must follow a constant-radius path for the relatively short arc, i.e., the length of the tire mark. (See also centrifugal force and centripetal force, Paras 8.017–8.022.)

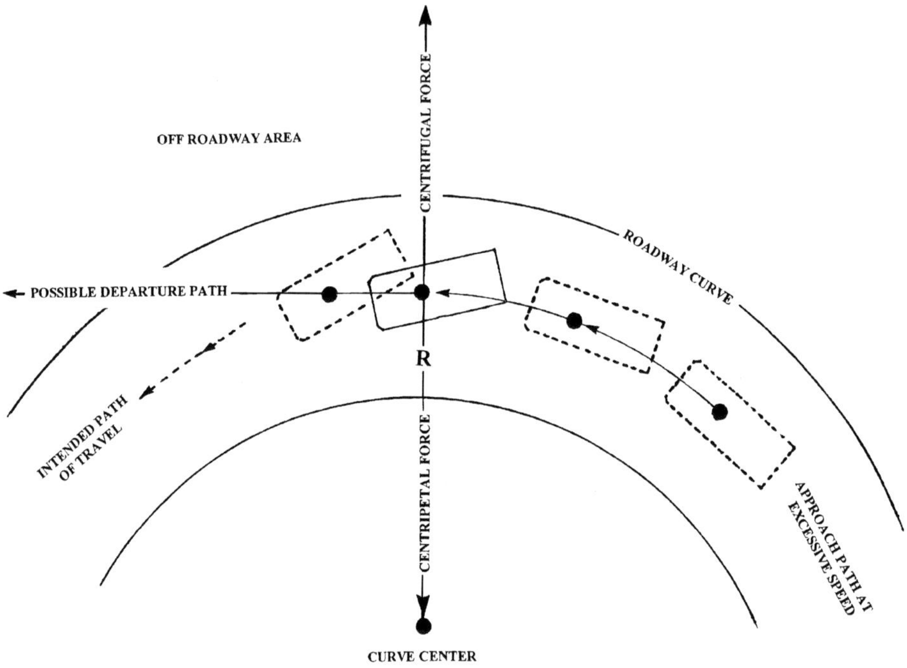


Figure 4–02. When a vehicle is steered to keep it in a curve, the roadway surface must furnish, through friction with the tires, a force that will keep it from traveling in a natural straight-ahead path. This force is called *centripetal* (toward the center) force. If the speed is too great, *inertia* (commonly known or referred to as centrifugal force) will overcome centripetal force and cause the vehicle to leave its intended curved path of travel and go straight ahead. In this example, *R* represents the radius of the curve path followed by the vehicle's center of mass.

### *Formula 4–01*

$$a_r = \frac{V^2}{R}$$

where $a_r$ = acceleration indicated by the change in velocity (acceleration rate)
$R$ = radius of the path followed by the vehicle
$V$ = the velocity of the vehicle

### *Example 1*

A vehicle travels at a constant speed of 120 mph or 176 ft/s [(193 Km/h or 53.6 m/s)] in a curve having a radius of 1,000 ft (304.8 m). The radial acceleration (because the vehicle is following a circular path) exhibited by the vehicle's center of gravity can be calculated as shown below. [Note: The tangential acceleration $(a_t)$ is equal to zero because the speed remains constant.]

| *English* | *SI* |
|---|---|
| $a_r = \frac{176^2}{1{,}000}$ | $a_r = \frac{53.6^2}{304.8}$ |
| $a_r = \frac{30976}{1{,}000}$ | $a_r = \frac{2872.96}{304.8}$ |
| $a_r = 30.976\ ft/s^2$ | $a_r = 9.43\ m/s^2$ |

**4.005** *Acceleration factor.* An acceleration factor is the ratio of an acceleration rate to acceleration due to gravity. For the purposes of this manual, acceleration factor is denoted by the symbol $f_a$. Using the result in *Example 1* as the acceleration rate, the acceleration factor can be calculated as follows:

***Formula 4–03***

$$f_a = \frac{a}{g}$$

where $f_a$ = acceleration factor
$a_r$ = acceleration rate
$g$ = acceleration due to gravity

| *English* | *SI* |
|---|---|
| $a_r = 30.976\ ft/s^2$ | $a_r = 9.43\ m/s^2$ |
| $f_a = \frac{30.976\ ft/s^2}{32.2\ ft/s^2}$ | $f_a = \frac{9.43\ m/s^2}{9.81\ m/s^2}$ |
| $f_a = 0.96$ | $f_a = 0.96$ |

***Example 2***

A vehicle that weighed 3200 lb (1451.49 kg) traveled at a speed of 60 mph or 88 ft/s [(96.5 km/h or 26.827 m/s)] around a curve that had a radius of 484 ft (147 m). The necessary centripetal force to hold it in the curve can be calculated as follows:

First, start with Newton's Second Law:

***Formula 4–03***

$$F = Ma$$

where $F$ = net force acting on mass M
$M$ = mass
$a$ = acceleration

This is a uniform force. Therefore, we apply Formula 4–01:

$$a_r = \frac{V2}{R}$$

where $a_r$ = acceleration rate
$V$ = velocity of vehicle in circular motion
$R$ = radius of curve

Next, insert the results in Formula 4–03

$$F = Ma$$

and we have:

***Formula 4–04***

$$F = \frac{MV2}{R}$$

where $F$ = centripetal force
$M$ = mass
$V$ = velocity
$R$ = radius

*Note:* In order to use this formula on the English side of this problem, we must change the weight in pounds into mass. As noted, this is done by dividing the weight (3200 lbs) by 32.2, the acceleration of gravity:

| *English* | *SI* |
| --- | --- |
| $F = \frac{MV^2}{R}$ | $F = \frac{MV^2}{R}$ |
| $F = \frac{3200}{32.2} \times \frac{88^2}{484}$ | $F = 1451.49 \times \frac{26.827^2}{147.523}$ |
| $F = \frac{3200}{32.2} \times \frac{7744}{484}$ | $F = 1451.49 \times 4.8785$ |
| $F = 99.37 \times 16$ | $F = 7081.06\ N$ |
| $F = 1589\ lb$ | |

# Chapter 5

# WORK

**5.001** *Work,* as used in physics, involves a unit of force multiplied by a unit of length, and is defined as:

> *A transference of energy from one body to another resulting in the motion or displacement of the body acted upon, expressed as the product of the force and the amount of displacement in the line of its action.*

If the force does not produce a displacement, no work is done in the sense in which work is used in physics.

**5.002** In the English system, when the force is measured in pounds and the distance is measured in feet, the work is expressed in foot-pounds (ft-lb). (Some sources use lb-ft to express torque and ft-lb to express energy in order to distinguish between the two.)

> *A force of one pound, acting through a distance of one foot, does one foot pound of work.*

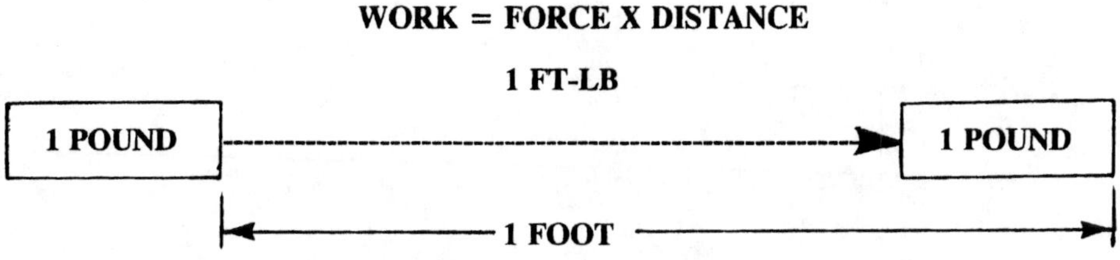


Figure 5–01A. A diagram depicting *work* in the English measurement system.

In SI, when the force is measured in newtons and the distance through which it acts is measured in meters, the work is expressed in joules (J):

> *A force of one newton, acting through a distance of one meter, does one joule(J) of work. (1J = 1Nm)*

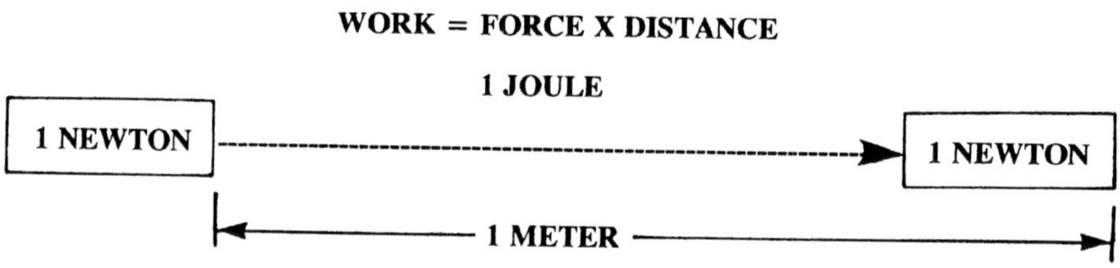


Figure 5–01B. A diagram depicting work in SI.

*Note:* The unit of measure used to describe the amount of work done is the *foot-pound* (ft-lb) in the English system and the *Newton-meter* or *joule* (N-m or J) in SI. One pound = 4.4482 newtons and 0.3048 meters 1 foot. Thus, 1 joule = .7376 ft-lb.

**5.003** In physics, work is synonymous with the term *energy transfer.* Work is done when a force *(F)* acts over a distance *(d)* in the direction of the force. In mechanics, work can be defined as the deformation experienced by an object when acted upon by a force. An example of work in effect is the expanding of an elastic band. In vehicle crashes, work is done in the bending and crushing of the metal. Technically, work is defined as the force multiplied by the distance through which the force is applied, explained algebraically as:

***Formula 5–01***

$$W = Fd$$

where $W$ = work done
$F$ = force that is in the same direction as $d$, or force applied
$d$ = displacement, or distance over which the force is applied in its direction of action

**5.004** *Displacement* is the movement of a body from one point or location to another with respect to direction as well as distance. It is therefore a vector quantity. A displacement is completely specified only when we state both its magnitude and direction. The magnitude consists of a number and a unit. For example, if a vehicle travels east 12 ft (3.6 m), strikes another vehicle and then travels 25 ft (7.6 m) northward to a final position, it has undergone a displacement of 12 ft (3.6 m) east and 25 ft (7.6 m) north (see Fig. 5–02). For traffic accident reconstruction purposes, we are primarily concerned with pre- and post-impact movements such as these. But its total displacement from the starting

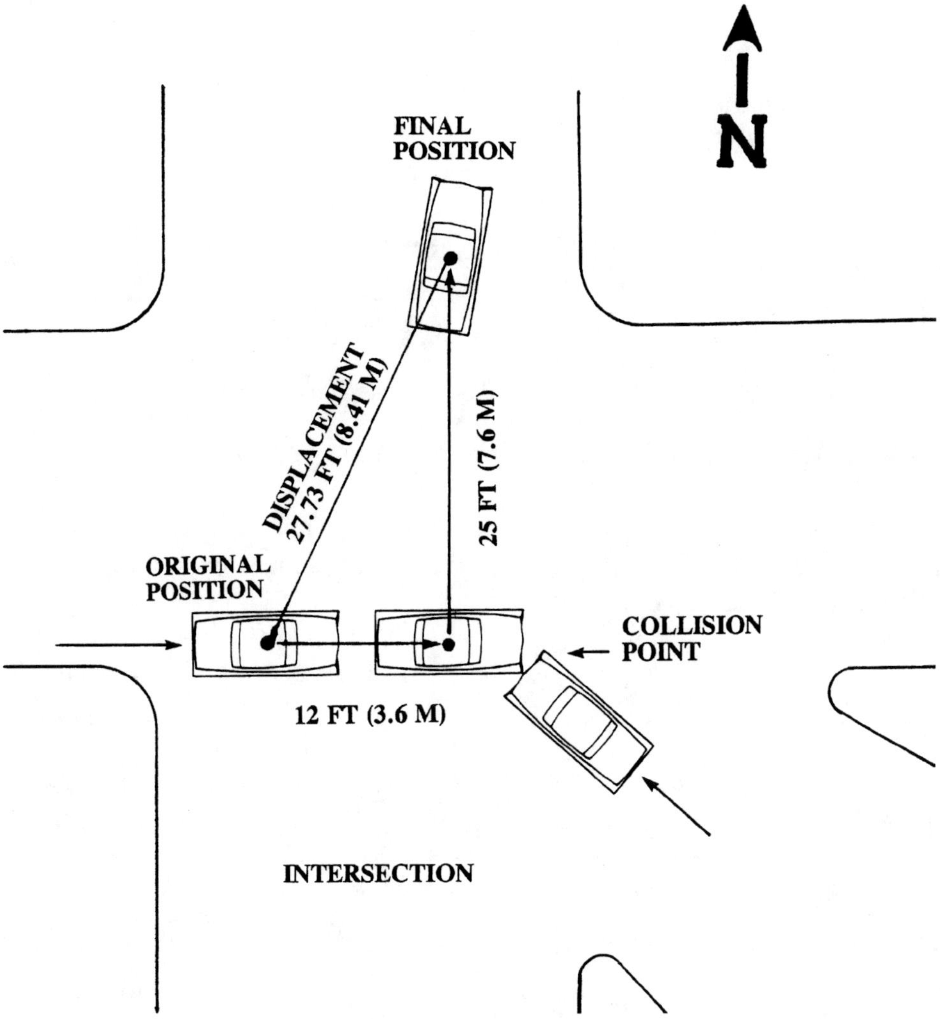


Figure 5–02. Displacement.

point is 27.73 ft (8.41 m) northeast, calculated as a trigonometric function of the angles involved. Formula 5–02 (Pythagorean theorem), which follows, is suitable for this calculation should it be desired, when the motion can be represented by a right triangle (a triangle with a right angle).

***Formula 5–02 (Pythagorean theorem)***

$$c = \sqrt{a^2 + b^2}$$

where $c$ = displacement or distance from starting point
$a$ = movement in the first direction
$b$ = movement in the second direction, perpendicular to the first direction

| *English* | *SI* |
|---|---|
| $c = \sqrt{12^2 + 25^2}$ | $c = \sqrt{3.6^2 + 7.6^2}$ |
| $c = \sqrt{144 + 625}$ | $c = \sqrt{12.96 + 57.76}$ |
| $c = \sqrt{769}$ | $c = \sqrt{70.72}$ |
| $c = 27.73$ *ft (northeast)* | $c = 8.41$ *m (northeast)* |

Because of the right triangle, in these cases, the angle of the vector can also be easily calculated using trigonometric relationships of *sin, cos,* or *tan.*

**5.005** When a force $F$ acts upon a body to produce a displacement, $d$, the work done by the force is defined as the product of the displacement and the component of the force in the direction of the displacement. If the angle between the displacement vector and the force vector is $\theta$ (see Fig. 5–03), the component of the force in the direction of the displacement is $F \cos \theta$, and the work done, $W$, can be determined by:

***Formula 5–03***

$$W = Fd \cos \theta$$

where $W =$ work
$F =$ force
$d =$ displacement
$\theta =$ angle

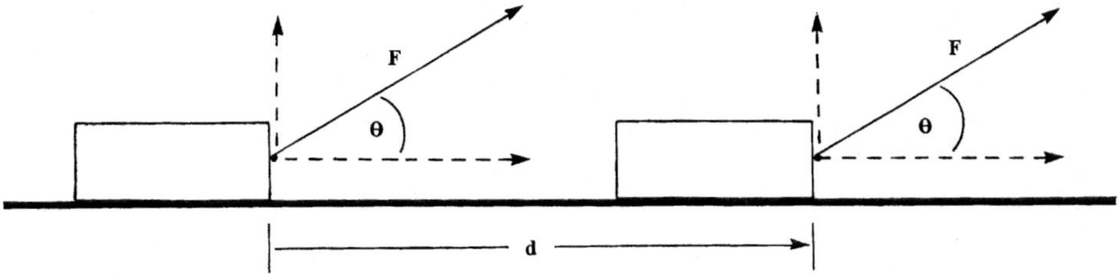


Figure 5–03.

***Example 1***

A rope is attached to a drag sled. An investigator holds the rope at an angle of 30° to the horizontal and uses a force of 10 lb (44.48 *N*) to move the drag sled 5 ft (1.524 m) across a level roadway. The work done to move the drag sled can be calculated as follows (see Fig. 5–03):

***Formula 5–03***

$$W = Fd \cos \theta$$

where $W =$ work in ft-lb *(J)*
$F =$ force in lb *(N)*
$d =$ distance in ft (m)
$\theta =$ angle ($\cos 30° = 0.8660$)

| *English* | *SI* |
|---|---|
| $W = 10 \times 5 \times 0.866$ | $W = 44.48 \times 1.524 \times .8660$ |
| $W = 43.3$ *ft-lb* | $W = 58.70 J$ |

*Note:* This method of pulling a drag sled is set out only for the purposes of explaining problem-solving when force (pull) is applied at an angle.

It is not the recommended method of pulling a drag sled. For the correct method, see Figure 8–01. For further explanation, see also *Drag Sled Concepts,* paragraph 9.013.

### *Example 2*

A 2,000 lb (907 kg = 8900 N) car skidded 85 ft (26 m) on a roadway having a drag factor of 0.75. The work exerted on the car can be calculated by multiplying the distance (d) by the force (F).

### *Formula 5–01*

$$W = Fd$$

where $W$ = work in ft-lb *(J)*
$F$ = force in lb *(N)*
$d$ = distance in ft (m)

| *English* | *SI* |
|---|---|
| $W = 85 \times (0.75 \times 2000\ lb)$ | $W = 26 \times (0.75 \times .8900\ N)$ |
| $W = 127{,}500\ ft\text{-}lb$ | $W = 172{,}857.9\ J$ |

# Chapter 6

# POWER

**6.001** Power is defined as *the time rate of doing work,* or *the amount of energy delivered per unit of time.* This can be set out as follows:

$$\textbf{POWER} = \frac{\textbf{WORK DONE}}{\textbf{TIME TAKEN}} = \textbf{FORCE X VELOCITY}$$

Figure 6–01.

*English*

**HORSEPOWER**

$$\text{POWER (in } hp\text{)} = \frac{\text{WORK (in ft-lb)}}{\text{550 ft-lb x t (in sec)}} \longrightarrow$$

Figure 6–02.

*SI*

**1 WATT**

1 NEWTON --→ ------------------------------------------------→ 1 NEWTON
--------------- 1 METER -------------→
------------ TIME = 1 SEC---------→

Figure 6–03.

In the English system, power may be expressed in foot-pounds per second (1 ft-lb/s). In SI, power is measured in joules per second or watts *(W)*. One watt = 1 joule per second (1 joule/s or 1 watt): (a) one watt of power is expended when one joule of work is done each second or (b) one watt of power is expended when a force of one newton acts through a distance of one meter in one second. Engine horsepower rating in the English system is defined as 550 ft-lb/s = 1 *hp*. In SI, horsepower is usually expressed in kilowatts (1,000 watts). Conversions are: 1 kW = 1.341 *hp*, and 1 *hp* = 550 ft-lb/s = 745.7 *W* (see Figs. 6–01, 6–02 and 6–03).

***Example***

An 80,000 lb (36,363 kg) tractor-trailer unit can develop 350 *hp* (261 *kw*) at the drive wheels. When traveling on a level roadway, assuming peak power at all times, and neglecting air resistance and shift times, the time that it would take the unit to accelerate to 50 mph (80.64 km/h) can be calculated using the following formula.[8]

***Formula 6–01***

*English*

$$P = \frac{WS^2}{30t}$$

*SI*

$$P = \frac{MS^2}{26t}$$

where $P$ = horsepower *(kW)*
$t$ = time
$M$ = mass
$S$ = speed
$W$ = weight
30 (26) = a constant

*Note:* The derivations of constants 30 and 26 can be found in: Daily, John, and Shigemura, Nathan S.: *Fundamentals of Applied Physics for Traffic Accident Investigators* (Vol. 1). Institute of Police Technology and Management, University of North Florida, Jacksonville, Florida, 1997.

We know from above that:

$$1\ hp = 550\ ft\text{-}lb/s \qquad 1\ kW = 1000\ watts$$

Therefore:

$$P = \frac{WS^2}{550 \times 30 \times t} \qquad P = \frac{MS^2}{26 \times 1000 \times t}$$

$$P = \frac{WS^2}{16{,}500t} \qquad P = \frac{MS^2}{26{,}000t}$$

Solve for *t:*

$$16{,}500t = \frac{WS^2}{P} \qquad 26{,}000t = \frac{MS^2}{P}$$

$$t = \frac{WS^2}{16{,}500\ P} \qquad t = \frac{MS^2}{26{,}000\ P}$$

$$t = \frac{80{,}000 \times 50^2}{16{,}500 \times 350} \qquad t = \frac{36{,}363 \times 80.64^2}{26{,}000 \times 261}$$

$$t = 34.6\ sec \qquad t = 34.8\ sec$$

**6.002** *Application of power to motor vehicles.* In mechanical engineering and traffic crash reconstruction, power is most often measured as *horsepower (hp).* The power that can be delivered by a motor vehicle engine is usually determined by its horsepower rating. This power is generated by the engine's revolutions per minute (rpm) and the torque (see Para. 6.003) exerted through the drive train to the wheels.

**6.003** A *torque* is a *twisting* or *rotary action;* a *force's turning effect* (or moment) *on an object.* The moment of a force (its torque) equals the force multiplied by the length of the moment arm on which it acts. The length of the moment arm is measured perpendicularly from the pivot point to the line of action of the force. The pivot point serves as the *center of moments,* the point from which the lengths of all moment arms are measured. Torque is commonly taken as positive if it tends to cause counterclockwise rotation and negative if it tends to cause clockwise rotation. A torque unit is a derived unit. In the English system, it is the *pound-foot (lb-ft)*; in SI, it is the *newton-meter (Nm)*. It should be noted that the terminology used for torque, i.e., lb-ft or Nm, differentiates it from energy (ft-lb or Joule). Torque can be shown in mathematical form in both systems as:

***Formula 6–02***

$$T = F \times L$$

where $T$ = torque
$F$ = force
$L$ = length of lever arm or moment

*Note:* See also paragraphs 11.011–11.014.

# Chapter 7

# ENERGY

## DEFINITION

**7.001** *Energy (E)* is defined as the ability or capacity to do work. In mechanics, we are concerned primarily with two forms of energy:

1. Potential energy *(Ep)*–see Para. 7.003
2. Kinetic energy *(Ke)*–see Para. 7.004

**7.002** *Energy,* like work, is a scalar quantity, i.e., it has magnitude but no direction (see also Chapter 5). Work units are used for measuring both potential and kinetic energy. The formula for calculating work when the force acts in the same direction as the displacement, is:

***Formula 7–01***

$$W = Fd$$

where $W =$ work
$F =$ force
$d =$ distance

## POTENTIAL ENERGY

**7.003** Potential energy *($E_p$)* is stored energy or the energy that a body (or system of bodies) has by virtue of its position and/or configuration.

Of particular interest is what is known as *gravitational potential energy.* This term is applicable where the potential energy of a mass, *M,* is predicated upon its height, *h,* in relation to some zero point. Algebraically, this can be shown as:

***Formula 7-02***

$$E_p = Mgh$$

where $E_p$ = potential energy
$M$ = mass in slugs
$g$ = acceleration due to gravity
$h$ = height

An example of an object possessing this kind of energy would be a vehicle sitting on an incline with the potential (potential energy) to roll downhill. In such a case, the vehicle has no potential energy relative to the roadway-tire contact area, but it does have potential energy relative to the roadway incline. If the brake fails to hold the vehicle stationary and it rolls, the potential energy is then transformed into kinetic energy. As the vehicle descends the slope, the potential energy decreases. Some of that may be converted into heat in the tire sidewalls or brake components as a result of the friction caused by the motion. Another example of an object having potential energy is a vehicle that crashes, when much of its kinetic energy may be converted into *crush* as strain, or deformation, energy.

## KINETIC ENERGY

7.004 *Kinetic energy* is defined as: *The energy a body possesses due to or by virtue of its motion.* To determine the kinetic energy possessed by a vehicle in motion, it is necessary to consider the *work* done on the vehicle in order to give the vehicle its speed. When the vehicle is brought to a stop, the energy is dissipated. An example of this is a speed determination based on the length of the skid marks which, in fact, is a measured quantity of energy for the moving vehicle. By definition, this is its kinetic energy which is expressed algebraically as:

***Formula 7–03***

$$Ke = {}^{1}/{}_{2}MV^2$$

where $Ke$ = kinetic energy
$M$ = mass
$V$ = velocity

**7.005** In the English system, kinetic energy is measured in foot-pounds (or equivalent slug-$ft^2/s^2$) if $M$ is in slugs and $V$ is in ft/s. In the SI, it is in joules (or equivalent kg-$m^2/s^2$) if $M$ is in kg and $V$ is in m/s or newton-meters.

## CONSERVATION OF ENERGY

**7.006** *Conservation of energy.* The principle of energy conservation is: *Energy cannot be created or destroyed.* It can, however, be transformed from one form of energy to another. In terms of vehicle collisions, when two vehicles collide, the laws of conservation of momentum (see Para. 10.003) and conservation of energy are both applicable, but in some collisions, part of the kinetic energy of the bodies is transformed into some nonkinetic form of energy such as heat or sound. Since the energy of a body is measured in terms of the work it can do, it has the same units as work. However, in the application of the law of conservation of energy to vehicle collisions, the problem becomes exceedingly difficult because many kinds of energy may be involved, some of which are difficult to measure. An in-depth explanation of this topic is beyond the scope of this manual and will not be dealt with further.

# Chapter 8

# FORCE

## DEFINITION

**8.001** Force is defined as:

a. *That which produces or prevents motion, or has a tendency to do so.*

b. *That which can impose a change of velocity on a material body, its direction of motion, or both, as explained by Newton's First Law of Motion.*

Force is often defined with respect to inertia (meaning idleness or laziness). The principle of inertia requires that in the absence of a force, a body will maintain its speed and direction (see also Paras 1.006, 8.016, 8.017).

**8.002** Force can be directed in two different ways:

1. Push
2. Pull

and its magnitude considered in two quantities:

1. Mass measured in lb or kg
2. Acceleration measured in ft/sec$^2$ or m/s$^2$

## FORCE CONCEPT

**8.003** To visualize the concept of force, consider a car standing on a roadway surface that has a coefficient of friction of 1.00. If the vehicle with all its wheels locked is placed in motion by being towed by a wrecker, the effort exerted by the wrecker to keep the car in continuous motion would be opposed by and equal to the force created by gravity acting on the car's mass. Regardless of the car's mass, we could thus say that the amount of force required to keep the car moving is equal to the vehicle's weight. This can be shown algebraically as follows:

***Formula 8–01***

$$F = Ma$$

where $F$ = force measured in lb (Newtons)
$M$ = mass in slugs (kilograms)
$a$ = acceleration in ft/s$^2$ (m/s$^2$)

**8.004** A muscular or mechanical exertion must be used to produce motion, or to extinguish it–the mechanical force of friction is used to brake a moving vehicle to a stop. A body that falls over a cliff is attracted downward by *gravitational force* or the force of gravity. If a sufficient force acts on a stationary body, it will cause the body to move or accelerate. If the body is already moving, the force may cause it to move faster (accelerate). If the direction of the force is opposite to the direction of the body's motion, the force will cause the body to slow down (decelerate), stop, or even reverse direction. The direction of the force is very important, and force must always be defined both by its size (magnitude) and its direction. This makes force a vector *quantity.*

**8.005** In order for a force to exist, something must push on something else. However, that something else pushes back on that something in equal measure. For example, it is commonly considered that when a vehicle collides with a fixed object the vehicle bounces back or away from the point of contact. This occurs because the fixed object exerts a force upon the vehicle equal in magnitude and opposite in direction to the force that the vehicle is exerting on it. These forces act simultaneously. The result is explained by Newton's Third Law of Motion, which states: *For every action, there is an equal and opposite reaction.* In reality, however, the bouncing back aspect of a collision is not a true example of this law because in some collisions–which are purely plastic–there will be no bounce back and yet the third laws still applies. The third law causes the vehicle to stop against the fixed object, then the characteristics of the deformed parts will determine the amount of rebound–again in conjunction with the third law. Another example is a vehicle traveling on a roadway. In this situation, the wheels of the vehicle push backward on the roadway surface, but at the same time the roadway surface pushes forward on the wheels with an equal force. Also, when a truck-tractor pulls a semitrailer along the roadway, the semitrailer pulls back on the truck-tractor with the same force as the truck-tractor pulls forward on the semitrailer.

**8.006** When one vehicle exerts a force on a second vehicle, as in a collision, the second vehicle exerts an equal and opposite force on the first, even if the first vehicle is a large truck that might be twice as massive as the second vehicle. It should be noted that when the masses of the two vehicles are multiplied, the car at half the truck's mass has just as much effect on the final product as the truck. This may be expressed mathematically as *ab,* where the car's mass = *a* and the truck's mass = *b.* If the car's mass *a* is multiplied by the truck's mass *b,* the product is *ab.* If the truck's mass *b* is multiplied by the car's mass *a,* the product is *ba,* which equals *ab.* If the car's mass is doubled, its mass becomes 2*a.* If the car's mass 2*a* is then multiplied by the truck's mass *b,* the product becomes 2*ab.* If the truck's mass is doubled, its mass becomes 2*ba,* which equals 2*ab.* It can be seen that by doubling either one of the two masses, the product doubles as well in either case and that the truck has no greater influence on this result than the car. This can be demonstrated more convincingly by substituting values for *a* and *b:*

| | | | | |
|---|---|---|---|---|
| *Car:* | *a = 1 ton* | *2a = 2 tons* | *2a x b = (2 x 1) x b* | *2 x 2 = 4* |
| *Truck:* | *b = 2 tons* | *2b = 4 tons* | *a x 2b = 1 x (2 x 2)* | *1 x 4 = 4* |

**8.007** *Force through distance concept.* To visualize the force through distance concept, assume we get a wrecker to cause a car of a certain weight to move one foot or one meter on a roadway that had a coefficient of friction of 1.00. To carry out the task, it would be necessary for the wrecker to use force measured in pounds or newtons equal to the weight of the car. If the car is moved two feet or two meters, the *work* (see Formula 8–02) done would be doubled. Similarly, if a crane lifts the same car one foot or one meter, the car has potential energy (to drop back to the earth) equal to its weight. If it is raised two feet or meters, its potential energy doubles, and so on.

***Formula 8–02***

$$W = Fd$$

where $W$ = work
$F$ = force
$d$ = distance

## UNITS OF FORCE

**8.008** One pound of force is defined as the amount of force that will produce an acceleration of 1 ft/sec$^2$ on a mass of 32.2 lb (a slug). One newton of force is defined as the amount of force which will produce acceleration of 1 m/s$^2$ on a mass of 1 kg (1 m/s$^2$ x 1 kg or 1 kg-m/s$^2$). In SI, the number very conveniently becomes *1* if we use newtons to measure the force, since we have already said that 1 newton gives 1 unit of acceleration to 1 unit of mass. So, the simple result is:

$$F\ (pounds) = M\ (lb) \times a\ (ft/s^2)$$
$$F\ (newtons) = M\ (kg) \times a\ (m/s^2)$$

where $F$ = force in pounds or newtons
$M$ = mass of object being moved, in slugs or kilograms
$a$ = acceleration

**8.009** The standard mass in the metric system is the *platinumiridium cyclinder* called the *standard kilogram.* By an act of the United States Congress, the standard pound is defined as having 0.45359237 times the mass of the standard kilogram (see also Para. 2.005).

## FOOT-POUND

**8.010** A *foot-pound* is defined as the work done by an applied force of one pound when it moves a mass through a distance of one foot in the direction of the force. In the English system, forces are measured in pounds, masses in slugs, and acceleration in feet per second per second. Converted to the corresponding SI unit, 1 pound-force = 4.448 222 *N*.

## THE NEWTON

**8.011** On a 1-kg mass, the force of gravity is 9.81 newtons. One newton (1 *N*) is that unbalanced force which will give a 1-kg mass an acceleration of 1 $m/s^2$, i.e.,a force of 1 kg-$m/s^2$. On a 2-kg mass, the force to get the same 1 $m/s^2$ is 2 x 9.81 or 19.62 newtons. On a 75-kg mass, the force is 75 x 9.81 newtons or 735 newtons, and so on. In short, the earth has a gravitational strength of 9.81 newtons per kg. It can be seen that 1 kilogram of weight does not equal one kilogram of mass, but rather it equals 9.81 *N*. For conversion purposes, 1 *N* = 0.224 809 pound-force (see also Paras 3.009–3.011).

## NORMAL FORCE

**8.012** The word *normal* is used to describe a line drawn perpendicular to another line or plane. The *normal force* on a surface resting (or sliding) on a second surface is the perpendicular component of the force exerted by the supporting surface on the surface being supported (see Fig. 8–01). Normal force, as it relates to the forces supporting an object in a gravitational field, is always directed at least partially upward, never downward.

*Note:* The term *normal force,* although not strictly needed in static situations, is being introduced here because it has been used traditionally in some of the equations throughout this manual and, more importantly, because you have to work with the concept of normal force in situations involving circular motion once you have advanced beyond the scope of this manual.

***Formula 8–03***

$$F = \mu N$$

where $F$ = force required to overcome friction force
$\mu$ = coefficient of kinetic (or sliding) friction.
$N$ = normal force opposing the perpendicular weight force/weight force component that presses the two surfaces together

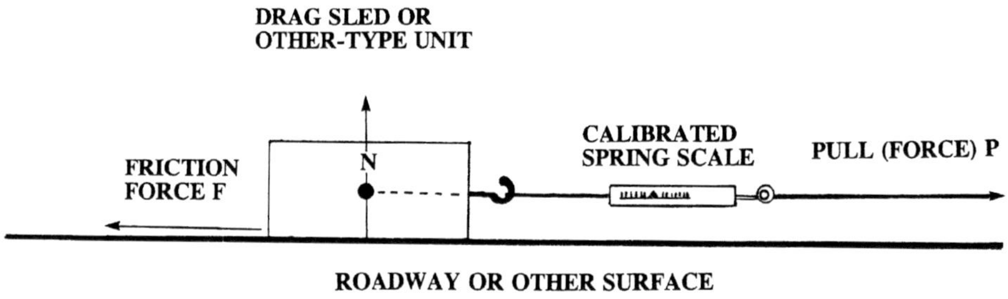


Figure 8–01A. Application of forces on a drag sled or similar-type device, where $N$ represents the normal force, $F$ the horizontal friction force, and P the friction force. Since the body is not accelerating downward, we know that the normal force equals the weight (W).

In the analysis of forces acting on bodies, it is common to use a *free-body* diagram which shows all the forces action on the body in question. For instance, given this situation, we can prepare the following:

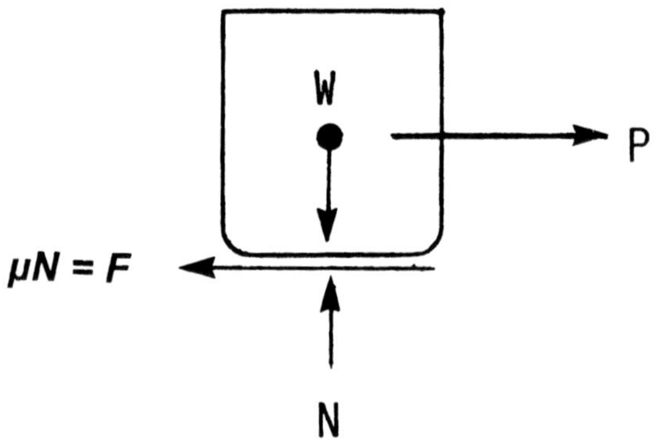


Figure 8–01B.

## FORCE, MASS AND WEIGHT

**8.013** The gravitational force acting upon each mass of a pair is directly proportional to the mass of each of them and inversely proportional to the square of the distance between their centers of mass. The value of $g$ is generally considered to be constant; therefore, weight may be said to be directly proportional to mass. Weight, $W$, is equal to the mass, $M$, of a body times the acceleration, $g$, due to gravity and can be expressed by Formula 8–04.

***Formula 8–04***

$$W = Mg$$

where $W$ = weight
$M$ = mass
$g$ = acceleration due to gravity

## FORCE WITHOUT MOTION

**8.014**

a. The foregoing paragraphs discuss the application of force against a body, giving the body acceleration. However, the question arises what happens, and how is it expressed, when a body, for whatever reason, cannot be caused through force to have motion? To answer this question, it must be realized that $F = Ma$ applies to a body that has no other forces acting upon it. If the body were in contact with another body, such as the ground, then the value $F$ in Newton's formula would be the vector sum of all the forces acting upon the body. If, for example, a vehicle, with all wheels locked up, were sitting on a roadway, Newton's $F$ would be the push exerted in trying to move the vehicle minus the force of friction keeping the vehicle in place. If the force exerted is insufficient to move the vehicle, then by Newton's Third Law, the friction force will be exactly equal and opposite; hence, Newton's $F$ will be *zero* and there will be no motion.

b. If the *push* on the vehicle is increased, a point will be reached where the friction force can be overcome. The friction force has a limiting value of the weight of the vehicle times the static coefficient of friction (see also Para. 9.004). Newton's *F* will now be the push exerted minus the weight of the vehicle as a resultant factor times the dynamic coefficient of friction. As long as Newton's *F* is positive, the vehicle will be accelerated. If the push is eased back until this force reaches zero, the vehicle will continue to move at a constant velocity.

c. Another thing that should be considered is that the push that is exerted on the vehicle will have the effect of deforming the vehicle, because all bodies exhibit some elasticity. The deformation will be a function of the push, not of Newton's *F.*

d. Consider a drag-sled test. With the sled stationary, the investigator exerts a pull on it to get it to move. Before it moves, the scale between your hand and the drag sled will show a reading of, say, 5 lb, 10 lb, 15 lb (or equivalent in kg) or more, as the pull is increased. Here the scale is the body of interest, and one can see that it becomes deformed as the pull is increased. Its deformation is what gives the reading (see Fig. 8–01).

e. Let us assume that when a reading of 20 lb (or equivalent SI) on the scale is reached, the sled keeps moving at a constant velocity. You can then get the value for the dynamic coefficient of friction. However, a yank or jerk on the scale might give a temporary reading up to 50 lb. The effect is that the sled will be accelerated proportionately.

## FORCE AND ACCELERATION

**8.015** Acceleration has the dimensions $ft/s^2$ ($m/s^2$), which can be read as *feet per second squared or feet per second per second ( meters per second squared or meters per second per second).* With the units for both mass and acceleration chosen, we can measure force with an appropriate unit. By Newton's second law, we have:

***Formula 8–01***

$$F = Ma$$

where $F$ = force in pounds (newtons)
$M$ = mass in slugs (kilograms)
$a$ = acceleration in ft/s$^2$ (m/s$^2$)

## CENTRIPETAL AND CENTRIFUGAL FORCES

**8.016** Before a discussion of centripetal and centrifugal forces takes place, the investigator should have an understanding of a number of key terms (see also Fig. 4–02):

***Centripetal.***

(a) Center-seeking; (b) directed, tending, or drawing toward a center; (c) pointing to the center of a circle.

***Centripetal Force***

(a) A force attracting a body toward a center around which it revolves; (b) a force that is required to hold a moving object in a circular path, such as a curve in a highway.

When an object is accelerated toward the center of a circle, the net force in that direction is called the centripetal force. When a vehicle travels in a curve at a constant speed, its velocity does not remain constant because it is continually changing in direction. It is centripetal force that causes this change in velocity.

***Centripetal acceleration***

Acceleration directed toward the center of a circle.

### *Centrifugal*

(a) Directed or tending away from a center; (b) center-fleeing.

### *Centrifugal force*

(a) The inertial reaction by which a body tends to move away from the center of rotation; (b) force that is the reaction to, or equal and opposite to centripetal; (c) an object's inertial resistance to being caused to move in a curved path.

In purely scientific terms, centrifugal force is an imaginary or apparent force, best explained by the term *inertia:* An object's inertia causes it to *want* to go straight by being pulled inward and forced to travel in a circular path, the inertia *feels* like it is pulling the object away from the center of travel (see Paras 8.017 & 8.021).

### *Equilibrium.*

(a) A body that remains at rest or moves with constant velocity is said to be in a state of equilibrium (equal balance). This is true however only if there is no resultant force and there is no resultant torque [see also Para. 1.003(1)(i)]. The balancing-out of forces when the resultant of all of them is zero. (b) The state of rest or balance due to the equal action of opposing forces.

**8.017** *Centripetal force* and *centrifugal force* are two forces that act upon a vehicle attempting to turn in a certain radius, as when traveling around a curve. In such cases, it is the centripetal force supplied by the friction between the tires and the road surface (and determined by the type of surface and any superelevation present) that holds the vehicle on the roadway and allows it to follow the curve. In the absence of this friction (centripetal force), the vehicle would travel in a straight line and leave the roadway tangent to the curve. Centrifugal force in such a case is the counteracting force whose magnitude is determined by the mass of the vehicle, its velocity, and the radius of the curve. When there is a *balance* or *equilibrium* between the two forces, the vehicle will travel along its intended path. However, when there is excessive speed for the curve and/or roadway conditions and equilibrium is lost, the vehicle will, depending upon its velocity, either travel in an arc of a radius greater than intended or in the extreme case of low friction travel a straight path tangent to the curve.

**8.018** When a vehicle travels around a curve, centripetal force acts inward on the vehicle. The vehicle itself is not in equilibrium, but is being continually accelerated in the direction of the centripetal force, i.e., toward the center of the curve. This may at first appear to be a paradox, for we have a vehicle moving with constant speed in a circle and yet being accelerated toward the center of the curve without getting any closer to it. If the adhesion between the tires and the roadway surface were to suddenly become nonexistent, however, the vehicle would fail to follow its intended path of travel around the curve, depart on a tangent and, by Newton's first law, travel in a straight line at a constant velocity. Another example of departure in this manner can be observed when water, mud or other debris is picked up by and adheres to a tire mounted on a wheel. When the wheel rotates more rapidly, this material may depart on a tangent once the centrifugal force exceeds the force of adhesion.

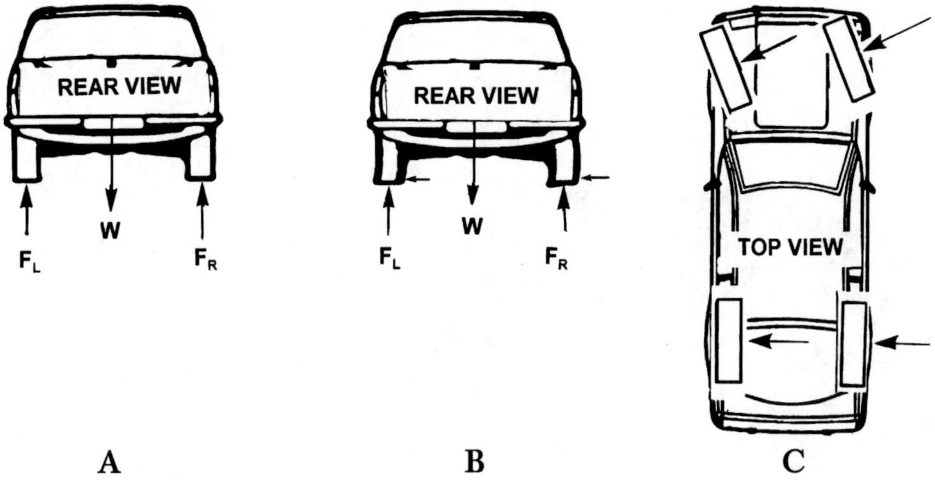


Figure 8-02. *Forces.* **A** shows a *rear view* of the front tires of a vehicle traveling in a straight path; **B** a rear view of the vehicle in a left-turn path of travel; **C** a top view showing the side forces on the vehicle's tires that hold it in a curved path.

**8.019** The acceleration described above is *centripetal acceleration* and is directed inward and toward the center of the circular path that the body is following. At any instant this acceleration is directed perpendicular to the velocity vector. The force associated with this acceleration is *centripetal force,* which holds the body in its path. The formula below relates the acceleration of the body to the radius of the path and the squared velocity of the body:

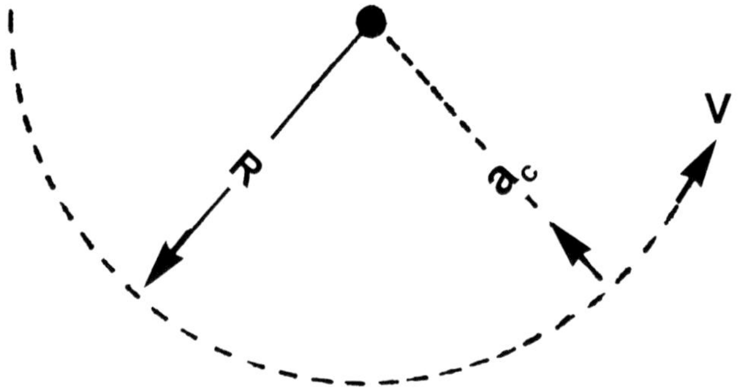


Figure 8–03.

***Formula 8–05***

$$a_c = \frac{V^2}{R}$$

where $a_c$ = centripetal acceleration
$V$ = tangential velocity of vehicle in circular motion
$R$ = radius of circular curve

Because $V = \omega R$, we also have $a_c = \omega^2 R$, where $\omega$ is in radian measure:

***Formula 8–06***

$$a_c = R\omega^2$$

where $a_c$ = centripetal acceleration
$R$ = radius
$\omega$ = angular velocity, $\frac{\text{radians}}{\text{sec}}$

*Note:* 1 radian = 57.296°. Knowing that there are $2\pi$ radians in a circle, we can multiply (degrees per second) by $\pi/180$ to get radians per second.

*Centripetal force* is the force that acts on the mass *M* moving in the circular path of its radius, *R*, to give it the centripetal acceleration, $V^2/R$. The force must be directed toward the center of the circular path. (When there is a balance or equilibrium between centrifugal and centripetal forces, the vehicle will travel along is intended path.) From the formula $F = ma$, we then have:

***Formula 8–07***

$$c_f = \frac{MV^2}{R}$$

or

$$c_f = \frac{W}{g} \cdot \frac{V^2}{R}$$

where $c_f$ = required centripetal force
$M$ = mass
$g$ = acceleration due to gravity
$W$ = weight
$R$ = radius
$V$ = velocity

## MAGNITUDE OF CENTRIPETAL FORCE

**8.020** When a vehicle travels in a curve in uniform circular motion, it experiences an acceleration toward the center of the circle, i.e., a centripetal force towards the center of the circle is produced. To find the magnitude of this centripetal force, apply Formula 8–05 ($a_c = V^2/R$), Formula 8–09 ($M = W/g$), and Formula 8–01 ($F = Ma$) as follows:

***Example***

A vehicle that weighs 3200 lb (1451.5 kg), travels on a level roadway at 60 mph (88 ft/s) [96.5 km/h (26.8 m/s)] around a curve that has a radius of 880 ft (268 m). The centripetal force required to keep the vehicle in the curve is calculated as follows:

Step 1: Calculate the centripetal acceleration ($a_c$).

***Formula 8–05***

$$a_c = \frac{V^2}{R}$$

where $a_c$ = centripetal acceleration
$V$ = tangential velocity of vehicle in circular motion
$R$ = radius of curve

| *English* | *SI* |
| --- | --- |
| $a_c = \frac{(88\ ft/s)^2}{880\ ft}$ | $a_c = \frac{(26.8\ m/s)^2}{268\ m}$ |
| $a_c = \frac{7744}{880}$ | $a_c = \frac{718.24}{268}$ |
| $a_c = 8.8\ ft/s^2$ | $a_c = 2.68\ m/s^2$ |

Step 2: Calculate the vehicle's mass ($M$).

***Formula 8–09***

$$M = \frac{W}{g}$$

where $M$ = mass
$W$ = weight
$g$ = acceleration due to gravity

*Note:* In the English system, the mass of a vehicle is calculated by dividing its weight by the acceleration due to gravity, i.e., by 32.2 ft/s/s. However, in the SI, there is no need to divide the weight by the acceleration due to gravity because its weight in kilograms is its mass.

| *English* | *SI* |
|---|---|
| $M = 3200/32.2$ | $M = 1451.5\ kg$ |
| $M = 99.37\ slugs\ or\ lb\text{-}s^2/ft$ | |

Step 3: Substitute ordinary force symbol $F$ for centripetal force symbol $c_f$ and then use once again Formula 8–01 to calculate the centripetal force:

***Formula 8–01***

$$F = Ma$$

where $F$ = force (in this case, centripetal force)
$M$ = mass
$a$ = acceleration

| *English* | *SI* |
|---|---|
| $F = 99.37 \times 8.8$ | $F = 1451.5 \times 2.68$ |
| $F = 874.5\ lb$ | $F = 3890\ N$ |

## CENTRIFUGAL FORCE AS A FICTITIOUS FORCE

**8.021** *Frames of Reference.* Martindale et al. (1986)[9] explain that all motion is relative and our description of it depends on the frame of reference used. It is common to use a fixed frame of reference. We consider the Earth to be such a frame of reference, and the motions of other objects are usually measured relative to it.

***Example 1***

If we throw a ball vertically upward inside a car that is moving at uniform velocity, we observe that the ball falls directly back into our hand. We get the same results whether the car is at rest or is moving at uniform velocity. This shows that while in the moving car, the ball was moving forward with its occupants at the same velocity, but appeared to just go up and down because the ball's forward velocity was the same as that of the occupants. In this situation, the frame of reference is the forward movement of the occupants. This is called a *moving frame of reference.* From this, and other similar observations, a generalization can be made: Newton's Laws of Motion hold equally well whether the frame of reference is moving at a uniform velocity or is at rest. This generalization is called the *Galilean Relativity Principle.*

***Example 2***

Your car is stopped alongside another car at a red light. It appears to you that your car is rolling backward, when in fact the other car is moving forward. Relative to the Earth (a *fixed frame of reference*), your vehicle remains at rest. What happens is that momentarily you use the other car as a frame of reference, assuming that it is the one at rest.

Since Newton's Laws of Motion concern the relationship between force, acceleration, and mass, they hold true in frames of reference at rest and in those moving at uniform velocity. These frames of reference are referred to as *Newtonian or inertial frames of reference.* However, frames of reference that are accelerating are not inertial frames, and Newton's laws do not hold.

***Example 3***

If a ball is placed on the floor of the box of a truck by someone standing in the box as the truck moves with uniform velocity, the ball will remain at rest on the floor. In this case, the truck is an *inertial frame of reference,* and the ball remains at rest relative to it. If, however, the truck accelerates in a straight line as the ball is set down, once the ball is released, it will roll straight but in the opposite direction to which the truck is heading. If the truck travels around a corner at uniform velocity, the ball will move sideways. In this case, the ball moves, and although it would appear that since there was a change in direction, there must have been a force on the ball to cause it to move. Yet there was none. In both the latter cases, the ball was observed in an *accelerating frame of reference.*

Discussions and disagreements often arise regarding the use of the term centrifugal force. It is a term that has had long-time use, and one that is commonly used in problem-solving when relating to centripetal force. Martindale et al. (1986) explain the term centrifugal force as an apparent or fictitious force as follows:

> Newton's laws do not apply in accelerating frames of reference. Since we are so familiar with inertial frames of reference, we are tempted to assume that there must be forces present whenever we observe accelerated motions, and we invent fictitious forces to explain these motions. A common example of such a fictitious force is the centrifugal force.
>
> When traveling in a car at a uniform speed around a curve, you suddenly feel that you are being pulled across the seat towards the outside of the car. You may assume that a centrifugal force is present, to explain this, since this would be the case in an inertial frame of reference. But you are making observations in an accelerating frame of reference, the car. What is actually happening is this: the frictional forces exerted on you by the car seat are not sufficient to provide the centripetal force necessary to keep you traveling in the same curve as the car. So you continue to move in a straight line, relative to the Earth, but slide across the seat, relative to the car, which is accelerating toward the center of the path of travel.
>
> To summarize, then, Newton's laws apply in inertial frames, but they do not apply in accelerated frames. Since the Earth is actually accelerating as it rotates about its axis, it is technically not an inertial system. However, the errors introduced by applying Newton's laws at the surface of the Earth are so small that we do not notice them experientially. For all practical purposes, therefore, we consider the Earth to be an inertial system, where the Newtonian laws of motion can be used.

*Note:* This explanation of centrifugal force as a fictitious force is provided for information purposes. The term centrifugal force, although not designating a real force, is an acceptable, relative term, widely recognized alongside inertia, where failure to generate adequate centripetal force is present. As is common practice in most traffic crash investigation and reconstruction textbooks, the term will be used in this manual.

## BANK OR SUPERELEVATION

**8.022** A vehicle may travel around a banked curve with its wheels pointed straight ahead and thus without any frictional force being generated to hold the vehicle in its circular path when the vertical component of the normal force is exactly equal to the vertical weight component of the vehicle and the horizontal component of the normal force is just equal to $MV^2/R$, the necessary centripetal force. The angle $\theta$ at which the road should be banked depends upon vehicle speed, not vehicle mass, and can be calculated using Formula 8–10 (see Fig. 8–04).

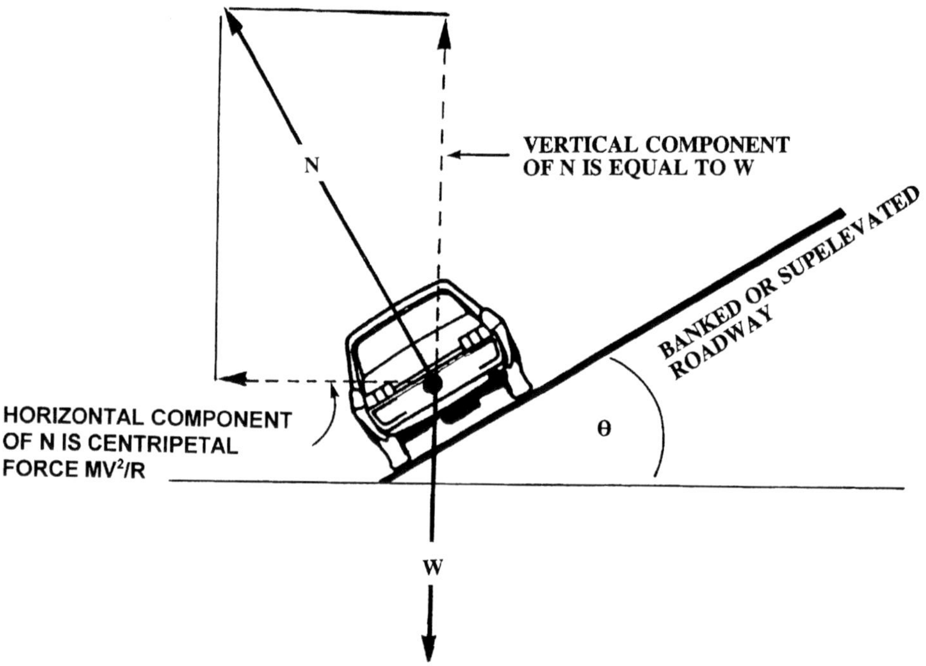


Figure 8–04.

***Formula 8–10***

$$\tan\theta = \frac{MV^2/R}{Mg}$$

$$\tan\theta = \frac{V^2}{gR}$$

where $\theta$ = tangent of the angle
$M$ = mass
$V$ = velocity
$R$ = radius
$g$ = acceleration due to gravity

# Chapter 9

# FRICTION

## DEFINITION

**9.001** *Friction* is defined as *a force tending to prevent motion between two surfaces that are in contact.* There can be no friction unless an attempt is made to cause one body to move relative to the other. When an attempt is made to move one body, a frictional force develops between them which tends to prevent or impede the motion, acting parallel to the surfaces that slide over each other, and in the direction opposite to that of the motion. Frictional forces are sometimes called *parasitic* because they only dissipate energy.

**9.002** Friction can be divided into two major categories:

1. *Static friction*–The friction between two surfaces in contact that are not moving relative to one another.

2. *Dynamic or kinetic friction*–The friction that is developed as a result of two surfaces that are in contact and sliding over each other.

## FRICTIONAL FORCE

**9.003** Frictional forces play an important role in mechanics, both for moving bodies and for those at rest.

a. When a body moves over a surface, there is always resistance to the motion. This resistance is known as frictional force.

b. Frictional force depends on many factors. For example, when a vehicle is driven along a roadway, the friction between the roadway surface and the vehicle depends on the composition and/or type of roadway surface involved and the composition of the vehicle part that is sliding or in motion over it. It could be a rubber tire or a piece of metal, as when a vehicle is sliding on its side. Also involved is the amount of force that is pressing the surfaces together. Frictional force does not depend on the gross area of the rubbing surfaces, but may depend on the pressure between the surfaces.

## COEFFICIENT OF STATIC FRICTION

**9.004** A body at rest on a surface has a weight, or weight component (if the surface is not level), that exerts a force at 90 degrees to the surface with which it is in contact. This force is resisted by an equal but opposite normal force $N$ (see Para. 8.012). The force of the weight/weight component presses together the contacting surface of the object and the surface on which the object is resting (see Figs. 8–01A, 8–01B, 9–01 & 9–02). The ratio of the force required to overcome this stationary condition (setting one body in motion with respect to the other) to that normal force is called the coefficient of static friction. The force needed to start the body moving, like the force then needed to keep the bodies sliding along one another at a constant speed, is a ratio of force to weight dependent on the magnitude of a frictional coefficient. Because the coefficient of static friction is greater than the coefficient of dynamic (sliding, kinetic) friction, the force needed to start the body sliding is greater than the force needed to keep it sliding once started. This starting force can be calculated with the following formula, which uses $N$ in place of $W$:

***Formula 9–01***

$$F_s = \mu_s N$$

where $F_s$ = force of static friction (and hence force needed to overcome static friction)
$\mu_s$ = coefficient of static friction.
$N$ = normal force (perpendicular force equivalent in magnitude but opposite in direction to the weight/weight component of an object in contact with a surface and acting at 90° to that surface).

## KINETIC FRICTION

**9.005** The *coefficient of kinetic friction* between two surfaces is *the ratio of the force required to pull one body along another to the normal force acting between the two bodies.* When a vehicle or other object slides over another surface, as when a tire skids on a roadway, the direction of the frictional force is parallel to the surface and opposite to the direction of the motion (see Fig. 9–01). The force necessary to overcome this friction is proportional to the normal force, and can be expressed algebraically as follows (see also Fig. 8–01):

***Formula 9-02***

$$F_k = \mu_k N$$

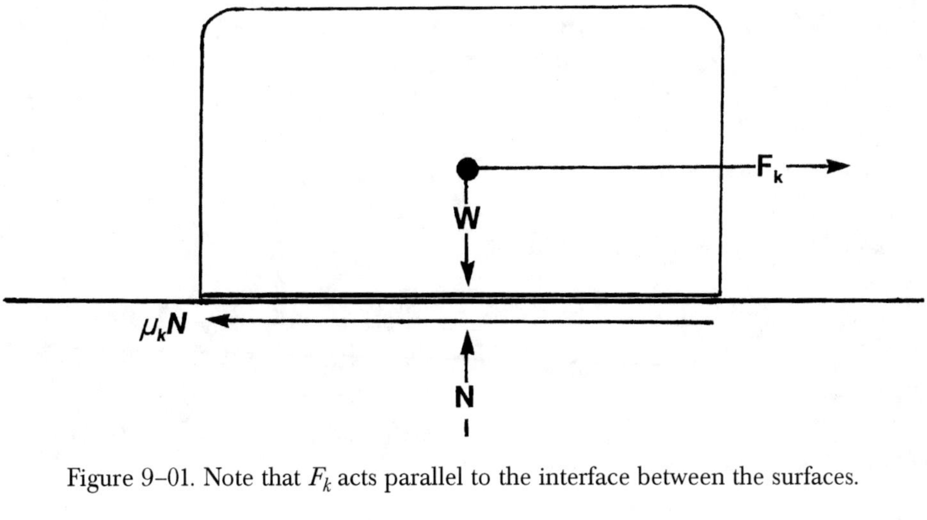


Figure 9–01. Note that $F_k$ acts parallel to the interface between the surfaces.

where $F_k$ = force necessary to overcome frictional force and keep a body moving with constant speed across a surface.
$N$ = normal force opposing the force pressing the surfaces together
$\mu_k$ = coefficient of kinetic, or sliding, friction (a constant)

## SLIDING AND ROLLING FRICTION

**9.006** In traffic crash investigation, examining the interaction between a tire and the roadway is of extreme importance. As a tire rolls, it utilizes static friction. As a tire slides, it utilizes kinetic friction. It can be said that rolling tires generate friction in two ways:

1. *Surface adhesion.* The microscopic chemical attraction between the tire material and the road material, and geometric interaction between the two surfaces.

2. *Hysteresis.* Irreversible energy loss due to the tire rubber conforming to small irregularities in the road surface, and due to flexing of the sidewall as it alternately gets loaded and unloaded during rotation.

**9.007** When we attempt to roll or slide one body over another, such as in the case of tires rolling or sliding over a roadway, we find that there is a force which opposes the motion. That force is called *friction.* It is the resisting force that has to be overcome if one body is to move or slide with respect to another body with which it is in contact. Ordinarily, frictional force is proportional to the load for a deformable object such as a tire, i.e., the *tire patch.* However, for a solid component such as a block of steel resting on a steel table, the contact area will not increase if the weight of the steel block is increased. Therefore, the contact area is not always proportional to the load.

**9.008** Frictional force is proportional to the load for *all* bodies, deformable or not. In the case of deformable bodies, e.g., tires, increasing the normal force can increase the contact area, but it has been found that for tires this relationship is not linear, due to carcass stiffness and edge-effects.

**9.009** Some physicists believe that friction is primarily caused by microscopic irregularities in the surfaces of the bodies. As the uneven surfaces are rubbed together, they tend to interlock and offer resistance to motion. The degree of friction differs not only among different materials, but also among different surfaces composed of the same material. Friction depends upon the materials and their surfaces and the forces holding them together.

**9.010** Apparently, friction arises primarily from adhesion of molecules of the two surfaces which are in contact. Even the smoothest surface is composed of hills and valleys from a microscopic point of view. Adhesion is not the same as geometric microscopic interlocking. The actual microscopic area of contact is generally only a tiny fraction of the macroscopic areas of the surfaces. Some substances melt and flow under the influence of pressure and frictional forces, particularly when high speeds are involved. Such behavior, called *plastic flow,* often lubricates the surface. For example: An ice skate slides effortlessly because it actually forms a very thin layer of melted water under the blade.

**9.011** Friction is independent of velocity and the force of friction between two surfaces may, therefore, be the same for a wide range of velocities. While there may be some deviations from this principle, its practical validity allows us to conduct test skids at lower speeds than the suspected speed of the vehicle involved in a collision, as well as to carry out tests with a drag sled. Other effects may, however, alter the relationships described here. That is why it is so important when conducting test skids that the conditions prevalent at the time of the accident be duplicated as closely as possible.

## COEFFICIENT OF FRICTION AND DRAG FACTOR

**9.012** *Coefficient of friction* and *drag factor* are perhaps the most commonly used friction terms in traffic crash investigation, coefficient of friction being designated by the Greek letter $\mu$ (pronounced "mew"), and drag factor symbolized with an italicized $f$. Both terms represent the ratio of stopping force to normal force expressed as a numerical value greater than zero and usually less than 1.1 or so for traffic accident reconstruction purposes.

## *Definitions*

a. *Coefficient of friction* ($\mu$) is defined as (see Fig. 8–01):

1. The sliding resistance between two surfaces when one of the surfaces is level, e.g., a roadway without grade or slope, and the other slides over that surface in a continuous motion.

2. The ratio of the force needed to overcome friction to the force pressing the surfaces together.

For traffic crash investigation purposes, coefficient of friction can be applied to a motor vehicle or to any other object, such as a part of a load or part of the vehicle itself, that is sliding over a level surface at a constant speed. In the case of a vehicle in a braking mode, if all wheels are skidding, the coefficient of friction will equal the kinetic friction coefficient.

b. *Drag factor* ($f$) is coefficient of friction adjusted for any and all matters that have an effect on the friction between two surfaces, such as, but not limited to:

1. Braking deficiencies on a motor vehicle that affect its ability to be braked to a stop, e.g., less than 100% braking capability or efficiency on one or more wheels of a vehicle, causing a wheel to not lock up.

2. A vehicle braking on a non-level surface, e.g., slope, grade or superelevation

To show algebraically the relationship between the factors involved in calculating the coefficient of friction or drag factor, i.e., the normal force $N$, the horizontal force $F$ and $\mu$, we may use the following formula:

### *Formula 9–03*

$$\mu = \frac{F}{N}$$

where $\mu$ = coefficient of friction (use symbol $f$ in the case of drag factor)

$F$ = force (pull), i.e., the pull required to cause continuous movement of an object across the surface

$N$ = normal force, i.e., force equal to the weight, $W$, of the object on a level surface, although acting in the opposite direction. See Para. 8.012.

## DRAG SLED CONCEPTS

**9.013** When a drag sled or other object is pulled across a *level* surface and the direction of the force (pull), $F$, is perfectly *horizontal* (parallel to the plane of the motion, i.e., perpendicular to the normal force between the object and the plane) as shown in Figure 9–01, the coefficient of friction, $\mu$, can be calculated by dividing the magnitude of the force, $F$, by the weight, $W$, of the object, i.e., $\mu = F/W$. It is important to observe here that if $F$ is not *parallel* to the ground, the normal force ($N$) pressing the two surfaces together is not equal to the weight of the sliding object. To move a drag sled across a surface, one possibility is to pull on an attachment such as shown in Figure 9–02A. We resolve the force $F$ into two components: the horizontal component $F_h$ is useful in overcoming the friction, while the vertical component $F_v$ reduces the force pressing the surfaces together. When the direction of the force (pull) is at an angle upward as shown in Figure 9–02A, the force $N$ pressing the surfaces together is reduced by the upward component $F_v$ of the applied force $F$ so that $N = W - F_v$. When the direction of the force (push in this case) is at an angle downward (see Fig. 9–02B), the downward component $F_v'$ of the applied force $F$ is increased so that $N = W + F_v'$. It can be seen therefore that when the force is applied at an angle, an adjustment must be made to compensate for the angle in order to calculate the coefficient of friction. The forgoing is for information purposes only, to illustrate the problems that can arise and the adjustments that must be made when an object is pulled along a surface other than in a manner described by Formula 9–03 and similar to that illustrated in Figure 8–01.

**9.014** Except in specific circumstances such as when instructions given by training institutions or mandated by departmental or prosecutorial

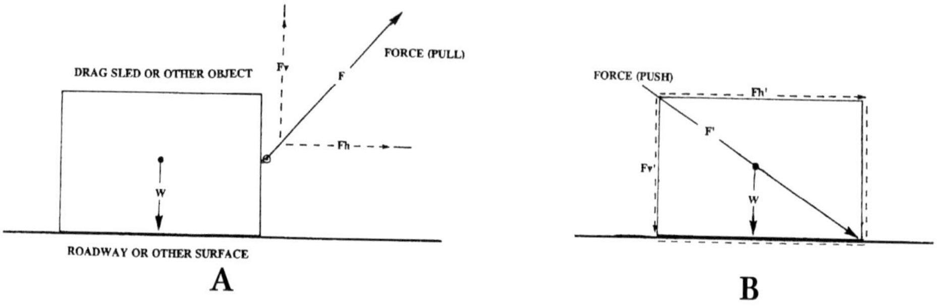


Figure 9–02A & B.

policies have dictated otherwise, it has been common practice in past years for traffic crash investigators to treat the terms *coefficient of friction* and *drag factor* as synonymous, but there are important differences between the two terms as they are used today. It is very important that the investigator understands these differences as presented in Paragraph 9.012 above.

## DERIVATION OF COEFFICIENT OF FRICTION

**9.015** The following is a method of deriving coefficient of friction:

***Formula 9–04***

**Coefficient of Friction**

$$\mu = \frac{F}{W}$$

Symbols used in the derivation:

$F$ = force
$M$ = mass
$W$ = weight
$g$ = acceleration due to gravity (32.2 fps/s or 9.81 m/s/s)
$\mu$ = coefficient of friction

Just as $W$ is a measure of the vertical acceleration of gravity, $F$ is a measure of a force exerted parallel to a surface.

Equivalents needed:

$W = Mg$

$F = Ma$

$a = gf$ or $g\mu$

### *Definitions*

*Mass* is the amount of matter in a body. Mass is equal to the weight of an object divided by the gravitational pull or acceleration. The gravitational pull on earth is considered *one g* and will cause an object in free fall to accelerate at a constant rate of 32.2 feet per second each second (32.2 fps/s), or 9.81 meters per second each second (9.81 m/s/s).

*Weight* is a measure of the action of the pull of gravity on the mass of an object, that is, the acceleration due to gravity times the mass. This can be stated as $W = Mg$.

Before beginning the derivation, you must first isolate $M$ in the following formula:

$$\boldsymbol{W = Mg}$$

Divide both sides by **g**.

$$\frac{1}{g} x \frac{W}{1} = \frac{1}{g} x \frac{Mg}{1}$$

Cancel terms.

$$\frac{1}{g} x \frac{W}{1} = \frac{1}{\not{g}} x \frac{M\not{g}}{1}$$

Rewrite.

$$\frac{W}{g} = M$$

Rearrange.

$$M = \frac{W}{g}$$

Now, start with the equation from Newton's Second Law of Motion.

$$F = Ma$$

Substitute $\frac{W}{g}$ for **M** and **g**$\mu$ for **a**.

$$F = \frac{W}{g} g\mu$$

Rewrite **F** and **g**$\mu$ as fractions.

$$\frac{F}{1} = \frac{W}{g} x \frac{g\mu}{1}$$

Cancel terms.

$$\frac{F}{1} = \frac{W}{\cancel{g}} x \frac{\cancel{g}\mu}{1}$$

Rewrite.

$$F = W\mu$$

Divide both sides by **W** (invert and multiply).

$$\frac{1}{W} x \frac{F}{1} = \frac{W\mu}{1} x \frac{1}{W}$$

Cancel terms.

$$\frac{1}{W} x \frac{F}{1} = \frac{\cancel{W}\mu}{1} x \frac{1}{\cancel{W}}$$

Rewrite.

$$\frac{F}{W} = \mu$$

Rearrange.

$$\mu = \frac{F}{W}$$

The formula can also be written as:

$$f = \frac{F}{W}$$

(Derivation source: Howell, Wiley L.: *Derivations Manual.* Institute of Police Technology and Management, University of North Florida, Jacksonville, Florida, 1994, pp. 69–72. Used with permission.)

# Chapter 10

# MOMENTUM

## DEFINITION

**10.001** Momentum, characterized by Newton's Three Laws of Motion, is defined as *the mass (or weight ) of an object multiplied by its velocity.* Thus, momentum, which we will symbolize with $M_{om}$, is the product of a scalar quantity (mass) and a vector quantity (velocity) and so is itself a vector quantity having direction as well as magnitude. It can be further stated that momentum is the property of an object in motion that tends to keep the object moving with the same velocity in the same direction. (Note: It is common is some textbooks to symbolize momentum with a lower case *p*.)

**10.002** In physics, when units are multiplied, they are expressed in hyphenated form or with an elevated dot between the variables. Thus, feet multiplied by pounds is expressed as ft-lb (*read* foot-pound–sometimes written lb.ft, for which read pound-feet). In the English system, momentum is expressed as slug-ft per second or slug-ft/s. This is because momentum is the product of mass (slugs) times velocity (ft/s). In SI, mass is expressed in kilograms and velocity is in meters per second. Thus, the unit of momentum is kg-m/sec. For conversion purposes, 1 pound-foot (i.e., 1 foot-pound) per second = 0.138 255 kg-m/s. Also, because $F \times \Delta t = M \times \Delta V$, momentum can be expressed in Newton-seconds. 1 Newton-second = 1 kg-m/s.

**10.003** Force changes the momentum of a vehicle or other free mass upon which it acts, and the rate at which it changes is directly proportional to the force that is applied. Accordingly, in terms of the Second Law, it can be said that:

> *The time rate of change of the momentum of a body is proportional to the net force acting upon the body and is in the direction of this resultant force.*

It must be noted, however, that the combined momentum of all the parts of a *system* (e.g., two colliding vehicles) does not change. In each case, the total momentum of the system is the same after as before the event.

**10.004** Knowing that acceleration is a change in velocity in a given change in time, we can rewrite Newton's Second Law (Formula 1–01A) as follows:

***Formula 10–01***

$$F = Ma$$

$$F = \frac{\Delta V}{\Delta t} = \frac{M\Delta V}{\Delta t}$$

where $F$ = force
$M$ = mass
$V$ = velocity
$a$ = acceleration
$t$ = time
(*Time:* see also Para. 2.004 and Formula 2–02A.)

**10.005** It can be assumed that in a practical sense, the force of gravity affects vehicles uniformly everywhere on the earth's surface. Also, for practical purposes in momentum, the terms *speed* and *velocity* can be considered interchangeable–provided we remember that direction of movement is implied when we use speed where normally we would use velocity. We can then treat momentum as speed *(S)* times weight *(W)* rather than in accordance with its formal definition as mass *(M)* times velocity *(V)*. The working formulae can then be:

***Formula 10–02A***

$$M_{om} = MV$$

where $M_{om}$ = momentum
$M$ = mass
$V$ = velocity (ft/s or M/s)

or

***Formula 10–02B***

$$M_{om} = WS$$

where $M_{om}$ = momentum
$W$ = weight
$S$ = speed (mph or km/h)

## IMPULSE

**10.006** *Impulse, I,* is the product of a constant or variable force *(F)* acting over a time period *(t)*. It generally refers to a very short period, such as the duration of an impact in a traffic collision, although it can refer to longer events as well. For example, the momentum of the space shuttle increases from zero during liftoff by the product of the engine thrust *(F)* and the burn time $(\Delta t)$. In terms of traffic accident reconstruction, the impulse results in a change of velocity of the body's mass *(M)* from say, zero $(V_o)$ at the start of the impact to its final velocity $(V_f)$ at the end of the impact. An understanding of this principle is very important in relating momentum to vehicle collisions

**10.007** The value of $\Delta t$ in a vehicle impact is very small, so much so the impulse on a system from external sources such as tires and wind can generally be neglected in these calculations. This assumption holds best during high-speed collisions where the change in momentum caused by such forces is small compared to the magnitude of the momentum of the system.

**10.008** Impulse units are units of force in pound-force (newtons) multiplied by units of time in seconds (recall that units of force are units of mass multiplied by units of acceleration) and can be calculated with the following formula:

***Formula 10–03***

$$I = F\Delta t$$

where $I$ = impulse [*lbf* x second ( newton seconds)]
$F$ = force
$t$ = time

**10.009** Impulse is the product of force and time that causes a change in the momentum of a vehicle or other object in collision. Since Newton's 3rd Law states that for every force there is an equal and opposite force, the impulse from one vehicle onto the other will be returned by the other vehicle as an equal and opposite impulse onto the first vehicle. Similarly, the resulting momentum change of each vehicle will be equal and opposite. Because these impulses are *internal* to the system, e.g., the two cars, they do not change the momentum of the system.

## CONSERVATION OF LINEAR MOMENTUM

**10.010** Linear momentum is the product of the mass and linear velocity. It is a vector quantity which has the direction of the velocity. The total momentum of vehicles or other objects in a system is not affected by their collision. The total momentum before collision equals the total momentum after collision, i.e., the momentum is conserved. Under this concept, it is possible to determine both the total momentum and the individual speeds of the vehicles at the moment of impact if other factors are known. For each vehicle, these factors include:

a. Mass or weight
b. Pre-impact direction of travel
c. Post-impact direction of travel
d. Post-impact speed, e.g., speed calculated from post-impact skid marks.

***Formula 10–04***

$$M_1V_1 + M_2V_2 = M_1V_3 + M_2V_4$$

where $M$ = mass
$V$ = velocity
$M_1V_1$ = Vehicle 1 (pre-impact)
$M_2V_2$ = Vehicle 2 (pre-impact)
$M_1V_3$ = Vehicle 1 (post-impact)
$M_2V_4$ = Vehicle 2 (post-impact)

This can be shown graphically as follows:

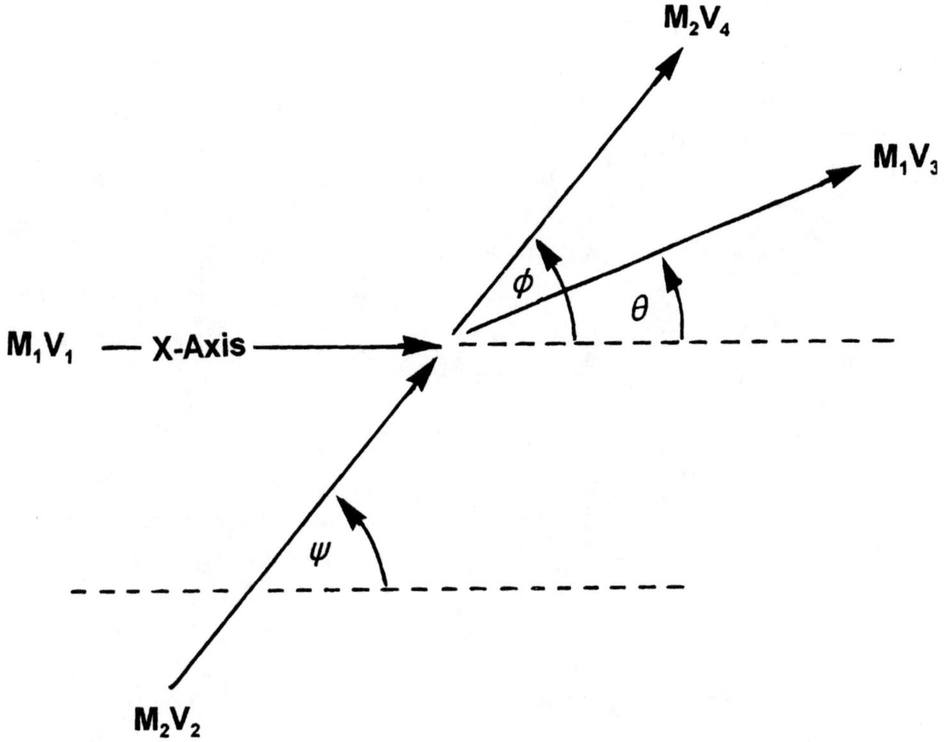


Linear momentum is conserved in both *x*- and *y*-directions. If one selects *x-axis* along $V_1$*'s* incoming path of travel, then the two linear momentum conservation equations become:

$$M_1V_1 + M_2V_2 \cos\psi = M_1V_3 \cos\theta + M_2V_4 \cos\phi$$

$$M_2V_2 \sin\psi = M_1V_3 \sin\theta + M_2V_4 \sin\phi$$

**10.011** Neglecting the surface forces that are present during the impact when a vehicle strikes another vehicle or other substantial movable object, and it breaks into sections or pieces, the resultant momenta of all the fragments after the collision, plus the momentum of the other vehicle or object after the collision, are equal to the sum of the pre-impact momenta of the two vehicles. Impact with immovable objects cannot be evaluated with this analysis method because the momentum transferred to, for example, a bridge abutment, cannot be measured.

**10.012** *Sensitivity.* Accurate measurements are crucial in making conservation of linear momentum calculations, particularly in respect to near head-on collisions. This includes the masses of the vehicles involved, departure velocities, and approach and departure angles. As pointed out by Daily (1988)*, most head-on collisions involve a small approach angle (5° or less). Seldom is it a true 180° opposite direction collision. If there is a small approach angle, 10° or less, conventional linear momentum analysis can be used. Because the approach angle changes rapidly for small angles, an accurately measured approach angle must be ensured. As an example, two vehicles are involved in a near head-on collision. The true approach angle is 5°. The departure angles are 10° for Vehicle 1 and 13° for Vehicle 2. Using these data, the collision speeds are calculated at 85 mph for Vehicle 2 and 58 mph for Vehicle 1. To show how sensitive angle measurements are, in this case if an error of 1° in the approach angle is made, such as rather than measuring a 5° angle, an angle of 4° is measured, the speed of Vehicle 2 is increased by 23 mph.

## MOMENTUM AND KINETIC ENERGY

**10.013** Momentum $(M_{om})$ and kinetic energy $(Ke)$ both are properties of moving objects. Whereas momentum is defined as the product of

*See Daily, J. *Fundamentals of Traffic Accident Reconstruction.* Institute of Police Technology and Management, University of North Florida, Jacksonville, Florida, 1988, pp. 233–235.

mass times velocity *($M_{om} = MV$)*, kinetic energy is defined as one-half the mass times velocity squared *($Ke = 1/2MV^2$)*. Thus, it is easy to see that as the speed of a vehicle increases, its kinetic energy increases faster than does its momentum. If the speed doubles, a vehicle's momentum doubles, but the *Ke* increases by a factor of four.

**10.014** When compared and worked out in examples, the formulae for momentum and kinetic energy show us that vehicles may have the same mount of momentum while having different amounts of kinetic energy. For example, if Vehicle 1 weighs 1,500 lb (or kg) and travels at 40 mph (or km/h), it will have the same amount of momentum as Vehicle 2 weighing 3,000 lb (or kg) and traveling 20 mph (or km/h). Vehicle 1, however, will have twice the kinetic energy of Vehicle 2.

**10.015** From the kinetic energy equation $Ke = 1/2MV^2$ (Formula 7–03), we see that kinetic energy can be quantified and there are times in motor vehicle crash investigation issues when this capability can be very useful. Energy is always conserved (see Para. 7.006), but in many circumstances, such as during a motor vehicle collision, the energy may change form or may be transferred. For example, as (a) when a vehicle slides to a stop, some of the kinetic energy possessed by the vehicle is converted into heat energy due to the friction between the tires and the roadway; and (b) when a vehicle strikes another vehicle or some other object, energy is used to deform the vehicles and possibly the passengers inside. In this case, all the kinetic energy is transformed into other forms of energy, being mostly *work* in the skid, but also a small amount to moving air, making noise, and heating the pavement.

## ELASTICITY

**10.016** In 1678, the English physicist Robert Hooke reported an important empirical relationship, since then referred to as Hooke's Law. This law states that *the deformation of an elastic body is directly proportional to the magnitude of the applied distorting force, provided the elastic limit is not closely approached or exceeded.* This law is valid for bodies composed of metals and many common structural materials. It is frequently not applicable to bodies composed of plastic materials. We refer to this relation as a law because of its limited range of application; however, in its generalized form, Hooke's Law constitutes the basis of modern theories of elasticity.[10]

**10.017** *Elasticity* is the property of matter that requires a force to produce distortion, and when the force is removed, causes matter to resume its original shape. An *elastic collision* can be defined as one in which no kinetic energy is lost as a result of the collision; the total kinetic energy after the collision is equal to the total kinetic energy before the collision. A *perfectly elastic collision* has a coefficient of restitution of 1.0 and that means there is no translational kinetic energy consumed in the impact. An object can be deformed during a collision, then regain its original form, but energy can still be consumed in the form of heat due to the deformation. A squash ball will regain its original form after being bounced off the floor, but it is obvious by the reduction in the rebound amplitude that energy was consumed in the collision with the floor. Cue balls that collide with very little distortion and a bounced rubber superball which regains its original shape are both examples of nearly perfect elastic collisions. If, however, there is major deformation caused to the object and the object fails to recover its original configuration, we say that the *elastic limit* of the object has been exceeded and that the object has acquired a *permanent set or permanent deformation.* Throwing a wad of chewing gum or silly putty against a wall is an example of a nearly perfect *plastic collision.* Whenever some of the total energy has been lost to the deformation, we consider it to be *inelastic.*

**10.018** Materials which can experience large deformations and still return to their original shape when released, are said to be highly elastic. Rubber bands and bungee cords are good examples of this. On the other hand, inelastic materials take a permanent *set* after even slight deformations. Ceramics and pottery are good examples of brittle materials. If they do not shatter during impact, they produce quite a high value of elasticity.

**10.019** In the real world, no collision is either perfectly *elastic* or *inelastic.* All collisions are actually somewhere in between, and are commonly called *semi-elastic* or partially elastic collisions. In vehicle collisions, *low-speed* impacts are at one extreme end of this spectrum: with very little energy dissipated in damaging the cars, twisting metal, and deforming bumper components, the total post-impact energy in the system is very nearly the same as the pre-impact energy. At the other end of the spectrum is the *high-speed barrier* collision with little to no rebound. This type of impact transforms all the incoming kinetic energy into crush energy, such that little or no kinetic energy is recovered by the vehicle.

**10.020** For purposes of traffic crash investigation, the scientific explanation of *elastic* and *inelastic* can be summarized by the criteria for the two basic classes of collisions: (1) If kinetic energy is conserved in the collision, the collision is classified as *elastic.* (2) If kinetic energy is not conserved, the collision is classified as *inelastic.* When two bodies stick together after collision, the collision is nearly inelastic. In a completely inelastic collision, however, all kinetic energy is not lost. Rather the loss of kinetic energy is consistent with the constraints imposed by the *conservation of linear momentum.*

## COEFFICIENT OF RESTITUTION

**10.021** Coefficient of restitution, symbolized with an epsilon, $\epsilon$, is a measure of elasticity, i.e., immediate after-collision rebound, spring or recovery from any deformation suffered during collision between two objects. The three types of collision described earlier can be further described by the *coefficient of restitution,* as follows:

1. In a perfectly elastic collision (where after collision, the object returns to its original dimensions or form (see also Para.'s 10.017–10.020), and the vehicles separate at a velocity that is relatively the same as their approach velocity, the coefficient of restitution is *1.*

2. In a perfectly inelastic (or plastic) collision (when two bodies stick together after collision (see also Para.'s 10.017–10.020), the coefficient of restitution is *0.*

3. In a semi-elastic collision (where the bodies spring apart after collision (see also Para. 10.019), the coefficient of restitution will be found to lie between the two extremes of *0* and *1.*

Coefficient of restitution for two vehicles can be shown algebraically using the following formula:

***Formula 10–05***

$$-(V_{1f} - V_{2f}) = \epsilon\ (V_{1i} - V_{2i})$$

where $V_f$ = final velocity
$V_i$ = initial velocity
$\epsilon$ = coefficient of restitution (A coefficient that lies between 0 and 1 )

# Chapter 11

# PRESSURE AND FORCE MULTIPLIERS

## DEFINITIONS

**11.001** *Pressure (p)* is defined as the force acting upon a surface per unit of area. In the English system, it is common to use *psi,* a colloquial expression for *pounds-force per square inch* where the pound is considered to be a unit of weight rather than a unit of mass. In SI, the units of pressure are *newtons per square meter* in the mks system, and *dynes per square centimeter* in the *cgs* system. If the force is measured in newtons and the area in $(\text{meter})^2$, then the pressure is in pascals (Pa) (see also Table 2–01.)

**11.002** *Pressure and stress.* Pressures and stresses are expressed in pascals (Pa), its multiples and sub-multiples. For instance, megapascal (MPa) is preferred over newton per square millimeter ($N/mm^2$). In terms of mechanical stress, properties such as tensile strength, compressive strength, flexural strength, and adhesion (shear method) are also expressed in pascals (Pa), its multiples and sub-multiples.

**11.003** Pressure can be calculated using Formula 11–01:

***Formula 11–01***

$$p = \frac{F}{A}$$

where $p$ = pressure

$F$ = normal force
$A$ = area

### *Example*

A 4-wheel, 2,000 lb car is at rest on a roadway surface. The surface area of each tire in contact with the roadway is 2 in x 5 in, or 10 square inches. The total contact area by all tires is therefore 4 x 10 = 40 square inches.

$$p = \frac{2000}{40}$$

$$p = 50\ psi$$

In the English system, it is common practice to refer to a pressure as so many *psi,* i.e., *pounds per square inch.* In SI, the name pascal (Pa) is used to express a pressure of one newton per square meter. For conversion purposes, 1 psi = 6.894 757 kPa.

*Note:* The kilogram-force (kgf) is not part of SI and is to be avoided in units of pressure such as the kilogram-force per square centimeter. Where kilogram-force per square centimeter ($kgf/cm^2$) is found used, it should be replaced by 100 kPa (an approximation).

**11.004** *Atmospheric Pressure.* In the English system, the normal pressure of the atmosphere at sea level is 14.7 pounds per square inch. This is known as a pressure of one atmosphere. Two atmosphere pressures is twice this, or 29.4 pounds per square inch. In SI, one standard atmosphere is 101.325 kPa and one technical atmosphere is 98.0665 kPa. However, in general use, it is considered to be 100 kPa at sea level. The actual value of air pressure varies from place-to-place and hour-to-hour, depending upon the weather. *Pneumatics* (from the Greek word air) is the term used in physics to describe the mechanical properties of air and other gases.

**11.005** *Absolute pressure* gives the actual pressure at a point in a fluid. Gauge pressure is the difference between the absolute pressure at a point in a fluid and atmospheric pressure. A pressure gauge measures pressure above or below atmospheric pressure. When it is necessary to differentiate between absolute and gauge pressures, the appropriate abbreviations should be added to the quantity symbol and not to the unit symbol.[11] Typical car tires have 35 psig (pounds per square inch gauge pressure) which is equivalent to approximately 49.7 psia (absolute).

**11.006** A common-type tire pressure gauge used to measure the pressure of a vehicle tire involves the motion of a spring-loaded piston in a cylinder. Air emerging from the tire causes a displacement of the piston against opposing forces exerted by the atmospheric air, friction, and a spring. It should be noted here that tire air pressure does not increase as the vehicle weight (load) increases. This is because as the load increases, the area of the tire patch in contact with the roadway surface also increases and as stated above, force = pressure x area.

**11.007** The laws of motion apply to fluids as well as to solids, and the study of mechanics includes in its broad sense, forces and motions involving fluids as well as solids. However, it is quite common to restrict the use of the term *mechanics* to solid bodies. It should be understood, however, that when we deal with fluids, we study both *hydrostatics* (the science that deals with the pressure and equilibrium of fluids, especially liquids) and *hydrodynamics* (the branch of dynamics that deals with the motions and forces of liquids, especially water).[12]

**11.008** An Irish scientist, Robert Boyle, found that:

> *If the temperature of the gas is kept constant, then the volume will be inversely proportional to the pressure.*

This became known as *Boyle's Law.* From this, it is evident that when the pressure is doubled, the volume becomes half as much; if the pressure is tripled, the volume becomes one-third of that it was, and so on. In the form of an equation:

***Formula 11–02***

A. $\frac{V_1}{V_2} = \frac{p_2}{p_1}$

or

B. $p_1V_1 = p_2V_2$

An inverse proportion is where $p_1$ and $V_1$ are, respectively, the pressure and the volume in one case and $p_2$ and $V_2$ are these values in another.

*Note:* The above formulae hold only at constant temperature. In general, compression of a gas will cause it to heat up and Boyle's Law does not apply under these circumstances. For example, a bicycle tire pump heats up when the tire is pumped up.

**11.009** If a constant supply of compressed air or fluid is directed through a pipe, it will flow out the other end as long as the supply lasts. If a plug were placed at the end of the pipe, the supply of compressed air or fluid would apply force against the plug. If a scale were to be placed against the plug, it would register just how much force was being exerted against the plug and, of course, against the walls of the pipe (see Fig. 11–01). If, for example, the plug had a surface area of 1 square inch (or 6.4516 $cm^2$) and the scale measured a force of 10 pounds (44.48 N), this force would be 10 psi (6.894 $N/cm^2$ or 68.947 kPa). It follows that if there was a pressure of 50 psi (344.7 kPa) in the pipe, the force exerted by the compressed air or fluid against the plug would be 50 psi (344.7 kPa) (see Fig. 11–02). For conversion purposes, 1 pound-force per square inch (psi) = 6.894 757 kPa.

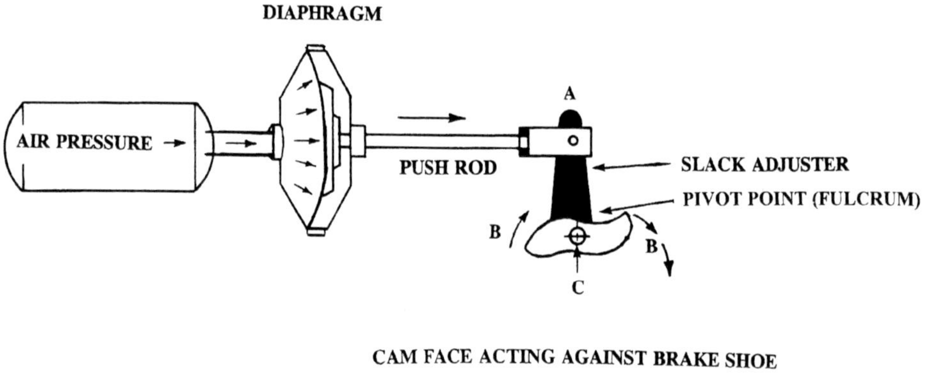


Figure 11–01.

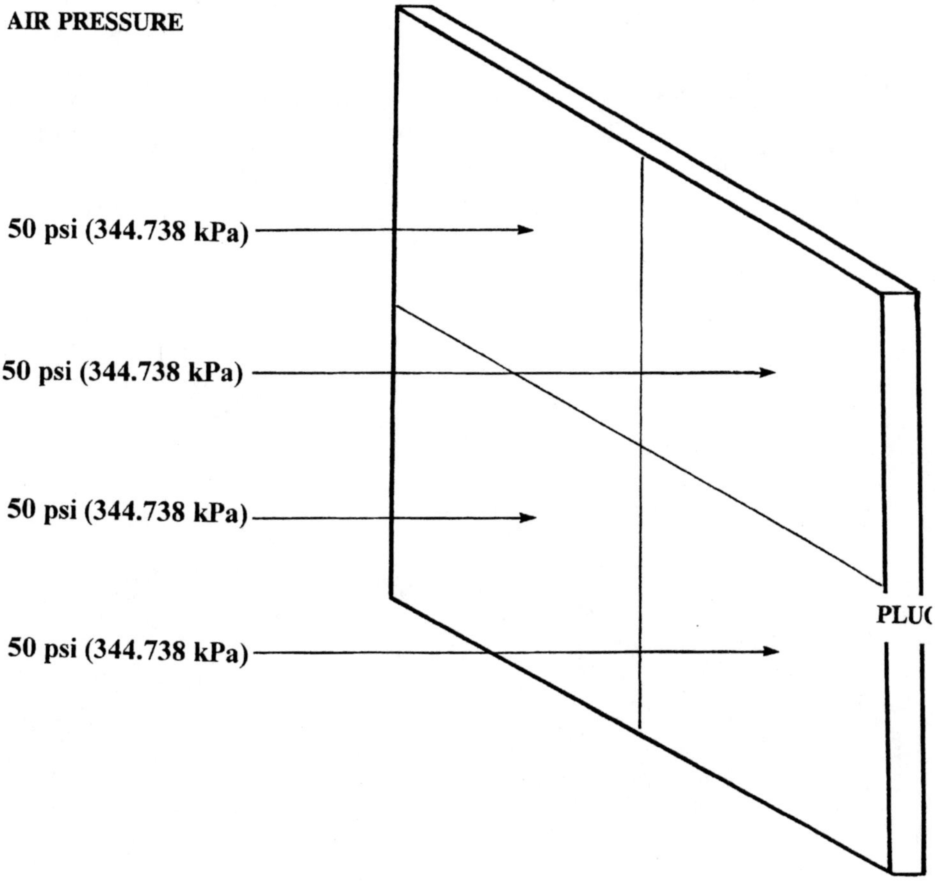


Figure 11–02.

**11.010** In actual practice, pipes are round and plugs are diaphragms of flexible material acting against push rods. If the structure of the chamber set out in Paragraph 11.009 and depicted in Figure 11–01 were changed so that the plug was 2 inches (or centimeters) on each side, the plug would have an area of 4 square inches (or square centimeters) (see Fig. 11–02). If the air pressure in the chamber maintained a constant 50 psi (344.738 kPa), there would be 50 pounds (220 N) force directed against each square inch (square centimeter) of the plug's surface, or a total of 200 pounds (880 N) force. In such a case, the walls of the chamber and the lead-in pipe would also experience the same pressure.

## FORCE MULTIPLIERS

**11.011** All types of braking systems use various forms of force multipliers, such as air, fluids, and mechanical leverage devices, often in various combinations. An example of mechanical advantage can best be illustrated by a simple lever system as shown in Figures 11–03 and 11–04.

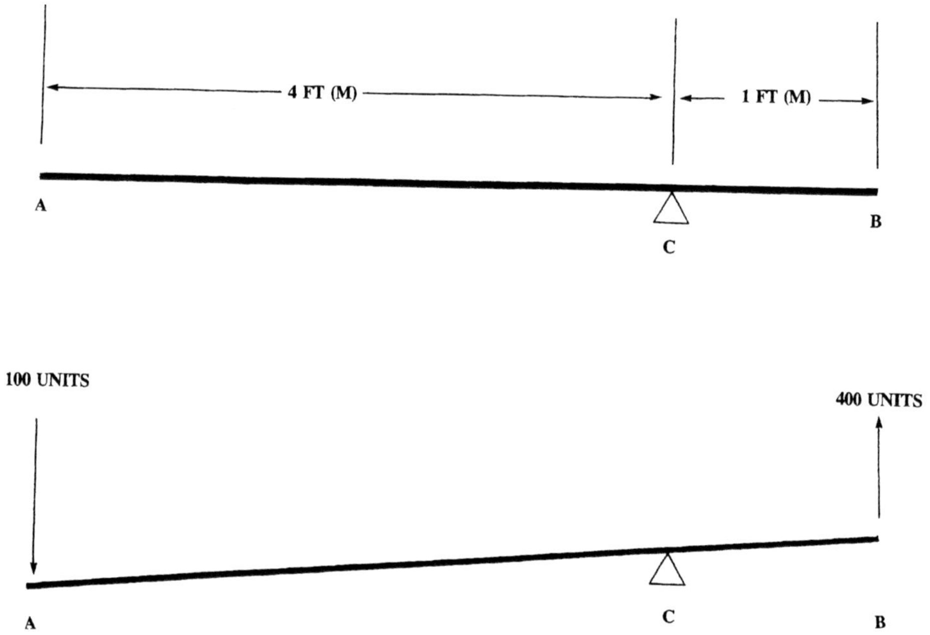


Figure 11–03. In the upper figure, a lever *(A-B)* is placed over a pivot point *C* (the fulcrum). The distance from *A* to *C* is 4 ft. The distance from *C* to *B* is 1 ft. This is a ratio of 4:1. If, as shown in the lower figure, a force of 100 lbs is applied down at point *A*, the upward force at point *B* will be 4 x 100 = 400 lb.

## AIR AND FLUIDS (LIQUIDS) AS FORCE MULTIPLIERS

**11.012** *Air.* Air is compressible. That is to say, a given amount (volume) of air can be compressed into a smaller space than it previously occupied and as such, can be a force multiplier. The smaller the space it is forced into, the greater resistance to being squeezed (pressure) it will have. An example is air compressed in the tires of the vehicle sup-

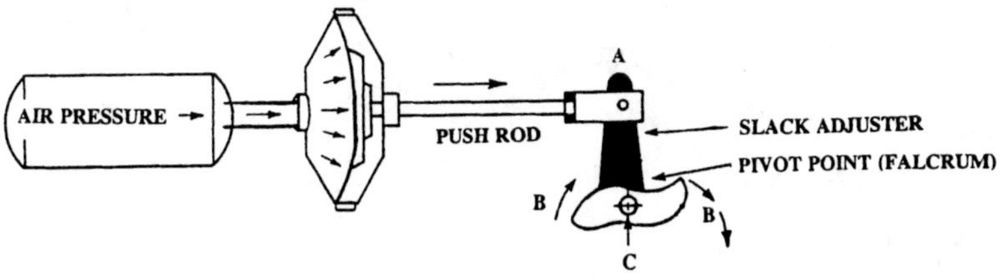


Figure 11–04. The lever example shown in Figure 11–03 can be applied to an air-brake system slack adjuster and cam by comparing points A, C, and B in the two figures.

porting the vehicle's weight. In the air brake system, this pressure is used to gain a mechanical advantage.

**11.013** *Fluids (Liquids).* If the object placed under pressure is a fluid, its volume is compressed, albeit only slightly as compared to gases. The compression is so slight that many fluids (including brake fluid and hydraulic oil) can be treated as incompressible. The concept of fluid force multipliers uses the fact that, apart from density effects, the pressure is the same throughout the fluid. Thus, applying a minimal force to a small area of the fluid can produce the same force over the entire area of the system where the fluid is pressing against the surfaces containing it. This is the principle of the motor vehicle hydraulic brake system, where pressure is applied mechanically by means of a brake pedal to a small piston to increase the amount of pressure on the wheel cylinder pistons (see Fig. 11–05).

## LEVERAGE AND AIR PRESSURE

11.014 Pipes used in braking systems are round and plugs, as used in heavy truck brake systems, are diaphragms of flexible material acting against push rods. If compressed air of 120 psi (827.37 kPa or 82.7 N/cm$^2$) acts on a *diaphragm* of 30 square inches (193.55 square centimeters), 3600 lb (16,007 N) of force is produced, i.e., 30 x 120 = 3,600 lb (82.7 x 193.55 = 16,007 N = 16kN). If this force is applied to a push rod in an air brake system to move a 6-inch (15.2 cm) slack adjuster operating a cam, a torque of 21,600 in-lb or 1800 ft-lb is gen-

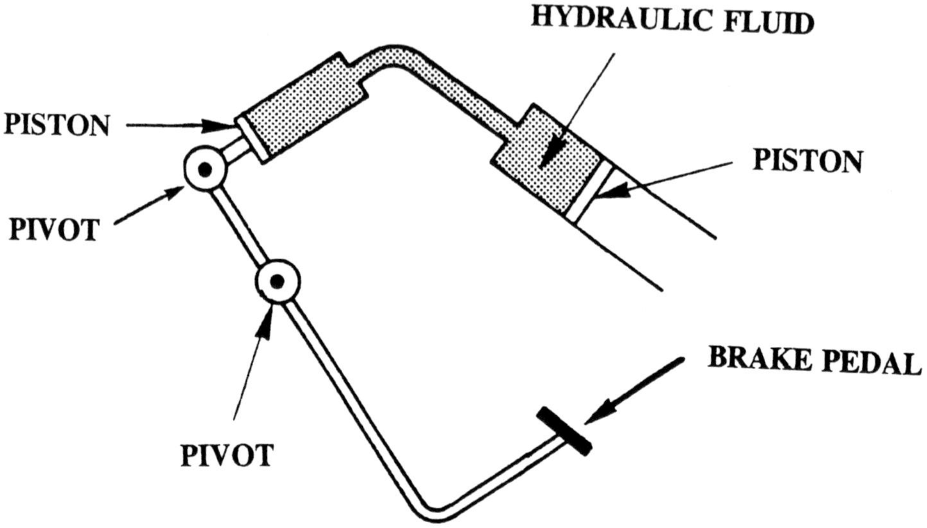


Figure 11–05. A common-type hydraulic brake system. A force is applied by the driver on the brake pedal at one end of a lever arm. The other end of the lever arm is connected to a piston. A hydraulic fluid fills the space between this piston and other pistons, the force being transmitted to them through the fluid.

erated. If this force is applied to a push rod 15.2 cm long, a torque of 16,007 N x 15.2 cm = 243,306 N·cm = 2433 N·m would be generated.

## TENSILE STRESS

**11.015** In the case of load restraining devices, such as tie-downs used in commercial vehicle operations, and tow-bars, forces tend to act outward to cause an increase in their length. This is known as *tensile stress.* When the forces act inward on the ends of an object, causing its length to decrease, there is a compression known as compressive stress. An example of *compressive stress* is in an object that supports weight. When an object is under too much stress, it may break or fracture. Each object has a maximum tensile force per unit area that it can withstand known as its *tensile strength,* the values of which can vary depending upon such factors as temperature and the molecular and crystalline structure of the material. Local legislation often sets out mandatory values of tensile strength for vehicle-related products. In the event of product failure, manufacturers are very often able to provide a tensile strength standard for their product, and many laboratories and engineering firms can provide an analysis of a failure that includes the forces required to cause such failure.

# APPENDIX A

## SYMBOLS

The following table provides a ready reference of symbols and definitions commonly used in traffic accident investigation and reconstruction, which will assist the reader in gaining a better understanding of the more complex formulae found in this manual, as well as those used in other traffic accident investigation and reconstruction textbooks. It should be noted that the symbols and definitions shown here may vary from those shown in other publications or used by various training sources. [Source: Rivers, R. W.: *Traffic Accident Investigators' and Reconstructionists' Book of Formulae and Tables* (2nd ed.) Charles C Thomas, Publisher, Springfield, Illinois, 1999, pp. 3–6.]

## GENERAL SYMBOLS, DEFINITIONS AND MEASUREMENT UNITS

| **Symbols and Descriptions** | **Measurement Units** | |
|---|---|---|
| | *U.S.* | *SI* |
| ***General*** | | |
| a = acceleration (±) | $ft/s^2$ | $m/s^2$ |
| C = chord | ft, in | m, cm |
| CG = center of gravity | ft, in | m, cm |
| CM = center of mass | ft, in | m, cm |
| d = distance, displacement | ft, in | m, cm |
| *e* = superelevation (See also lower case *m*) | percent | percent |

| | | |
|---|---|---|
| $f$ = drag factor | decimal fraction | decimal fraction |
| μ = coefficient of friction | decimal fraction | decimal fraction |
| f = acceleration factor (±) | Usually expressed as a decimal fraction | Usually expressed as a decimal fraction |
| $f_a$ = acceleration factor | Usually expressed as a decimal fraction | Usually expressed as a decimal fraction |
| $f_d$ = deceleration factor | Usually expressed as a decimal fraction | Usually expressed as a decimal fraction |
| $f_L$ = lateral acceleration factor | Usually expressed as a decimal fraction | Usually expressed as a decimal fraction |
| F = force in pounds (newtons) | lbf | N kgf |
| $g$ = acceleration due to gravity | 32.2 $ft/s^2$ | 9.81 $m/s^2$ |
| h = height | ft, in | m, cm |
| I = inertia | slugs | kg, grams |
| kg = kilograms | | |
| Ke = kinetic energy | | |
| L = length | ft, in | m, cm |
| l = length | ft, in | m, cm |
| M = mass | lb | kg |
| M = middle ordinate | ft, in | m, cm |
| $M_{om}$ = momentum | lb.ft/s | kg.m/s |
| P = power | ft-lb/s | joules or watts |
| $p$ = momentum | lb.ft/s | kg.m/s |
| $m$ = grade, slope (See also lower case *e*.) | percent | percent |
| $n$ = percentage factor | E.g., percent braking efficiency | E.g., percent braking efficiency |
| R = radius | ft, in | m, cm |
| r = run, rise or fall (Used in measuring grade, slope or superelevation) | ft, in | m, cm |
| S = speed | mph | km/h |
| t = time | hr, min, sec | hr, min, sec |

| | | |
|---|---|---|
| $t_w$ = track width | Usually expressed in inches | Usually express in cm |
| *Mu* = Coefficient of friction | Usually expressed as a decimal fraction | Usually expressed as a decimal fraction |
| (See also *f*, drag factor) | | |
| V = velocity | ft/s | m/s |
| W = weight | lb | kg |

***Symbols to Denote Specific Angles***

| | |
|---|---|
| $\alpha$ = *alpha* | Angle in degrees |
| $\theta$ = *theta* | Angle in degrees |
| $\phi$ = *phi* | Angle in degrees |
| ¥ = *psi* | Angle in degrees |
| $\Omega$ = *omega* | Any other specific angle |

***Common Symbols in Solving for Momentum Speed***

| | |
|---|---|
| $\alpha$ = *alpha* | Vehicle 1's approach angle |
| $\theta$ = *theta* | Vehicle 1's departure angle |
| $\phi$ = *phi* | Vehicle 2's departure angle |
| ¥ = *psi* | Vehicle 2's approach angle |
| $\Omega$ = *omega* | Any other specific angle |

***General Symbols in Traffic Accident Reconstruction***

*tan* = tangent

*tan* $\theta$ = tangent of angle $\theta$

$\in$ = *epsilon* (Used to symbolize coefficient of restitution)

$\Delta$ = *delta* (Used to indicate change, e.g., $\Delta V$ = change in velocity)

$\infty$ = directly proportional to

$\cong$ = approximately equal; congruent

$\sim$ = similar to; equivalent

$\infty$ = infinity

$\pi$ = *pi* (3.14159) The ratio of the circumference and a diameter of the same circle.

$\sqrt{}$ = radical sign

$\Sigma$ = *sigma*; summation of

$\pm$ = plus or minus

$\cdot$ = elevated dot, used as a multiplication symbol

*cos* = cosine

*sin* = sine

*tan* = tangent

$n$ = ad infinitum

$\therefore$ = therefore

$\vec{V}$ = An arrow above a variable to indicate that it is a vector

$W$ = work ft-lb N-m or j

$Y$ = Y axis; vertical axis

$X$ = X axis; horizontal axis

***Subscripts***

$X_o$ = subscript *o* denoting the *original* or *initial* factor for the component to which it is attached.

$X_1$ = subscript *1* denoting an *initial* factor, such as speed or velocity, or number *1* for the component to which it is attached when two or more factors are involved. Additional factors may be denoted by subscripts *2* and *3* and so on.

$X_f$ = subscript f denoting the *final* factor for the component to which it is attached.

***Overline***

$\overline{S}$ = average speed. The small bar (overline) denotes *average* for the component to which it is attached.

# APPENDIX B

## Table B–1

## Conversion Factors

### Linear Relationships

| *Length* | | *Km* | | *m* | | *ft* | | *mi* |
|---|---|---|---|---|---|---|---|---|
| 1 km | = | 1 | = | 1000 | = | 3280.83 | = | 0.62137 |
| 1 mile | = | 1.60935 | = | 1609.35 | = | 5280 | = | 1 |

### Velocity and Speed Relationships

| *Velocity* | | *m/s* | | *ft/s* | | *km/h* | | *mph* |
|---|---|---|---|---|---|---|---|---|
| 1 m/s | = | 1 | = | 3.282 | = | 3.6 | = | 2.240 |
| 1 ft/s | = | 0.30480 | = | 1 | = | 1.0973 | = | 0.6818 |
| 1 km/h | = | 0.27778 | = | 0.9113 | = | 1 | = | 0.6214 |
| 1 mph | = | 0.44704 | = | 1.46667 | = | 1.6093 | = | 1 |

### Volume

Gallon (U.S.) x 3.785 = liters

Gallon (Imperial) x 4.546 = liters

Liters x 0.264 = gallons (U.S.)

**Weight - Mass**

| | | |
|---|---|---|
| 1 short ton (2000 lb U.S.) | = | 0.9072 metric tonne |
| 1 long ton or 2240 lbs (Imperial) | = | 1.016 metric tonnes |
| 1 short ton or 2000 lbs (U.S.) | = | 907.18474 kg |
| 1 long ton or 2240 lbs (Imperial) | = | 1016.0469 kg |
| 1 pound (avoirdupois) | = | 0.45359 kg |

## Table B–2

**Speed of Light**

| *English* | *SI* |
|---|---|
| 186,300 miles per second | 300,000 kilometers per second |

## Table B–3

**Speed of Sound in Air**

| *English* | *SI* |
|---|---|
| 1087 ft/s | 331.5 m/s |

## Table B–4

**Acceleration Due to Gravity**

| *English* | *SI* |
|---|---|
| 32.174 ft/$s^2$ | 9.806 65 m/$s^2$ |

# REFERENCES


1. Williams, Dudley, & Spangler, John: *Physics for Science and Engineering.* D. Van Nostrand Company, New York, 1981, p. 35.
2. Smith, Alpheus W., & Cooper, John N.: *Elements of Physics* (9th ed.). McGraw-Hill Book Company, New York, (undated), p. 45.
3. Martindale, David G., Heath, Robert W., & Eastman, Philip C.: F*undamentals of Physics: A Senior Course.* D.C. Heath Canada Ltd., 1986, p. 3.
4. Martindale, David G., Heath, Robert W., & Eastman, Philip C.: *Fundamentals of Physics: A Senior Course.* D.C. Heath Canada Ltd., 1986, p. 3.
5. Rivers, R. W.: *Speed Analysis for Traffic Accident Investigation* (2nd ed.). Institute of Police Technology and Management, University of North Florida, Jacksonville, Florida, 1997, p. 116.
6. Bueche, Frederick J.: *College Physics* (7th ed.). McGraw-Hill Book Company, New York, 1979, p. 35.
7. Rivers, R. W.: *Speed Analysis for Traffic Accident Investigation* (2nd ed.). Institute of Police Technology and Management, University of North Florida, Jacksonville, Florida, 1997, p. 7.
8. Daily, J., & Shigemura, N. S.: *Fundamentals of Applied Physics for Traffic Accident Investigation* (Vol.1). Institute of Police Technology and Management, University of North Florida, Jacksonville, Florida, 1997, p. 224.
9. Martindale, David G., Heath, Robert W., & Eastman, Philip C.: *Fundamentals of Physics: A Senior Course.* D.C. Heath Canada Ltd., 1986, p. 189.
10. Williams, Dudley, & Spangler, John: *Physics for Science and Engineering.* D. Van Nostrand Company, New York, 1981, pp. 250, 251.
11. Williams, Dudley, & Spangler, John: *Physics for Science and Engineering.* D. Van Nostrand Company, New York, 1981, pp. 283, 284.
12. Asimov, Isaac: *Understanding Physics: Motion, Sound and Heat* (Vol. 1). A Mentor Book. New American Library, New York, 1966, p 120.

# BIBLIOGRAPHY

The following are sources referenced in developing this manual, and which are recommended for further reading and study.

Allen Corporation of America: *Physics and Measurement* (2nd ed.). U.S. Department of Transportation (NASS). Contract No. DTNH22-83-C-07023. (Undated).

Asimov, Isaac: *Understanding Physics: Motion, Sound and Heat.* A Mentor Book, New American Library. New York and Scarborough, Ontario, 1966.

Bueche, Frederick J.: *College Physics* (7th ed.). McGraw-Hill Book Company, New York, 1979.

Canadian Metric Practice Guide. Prepared by Canadian Standards Association, Standards Council of Canada (CAN3-Z234.1-76), Rexdale, Ontario, 1976.

Canadian Metric Guide. Council of Ministers of Education, Canada (2nd ed.). (ISBN 0–7744–0127–3), Rexdale, Ontario, 1976.

Daily, J.: *Fundamentals of Traffic Accident Reconstruction.* Institute of Police Technology and Management, University of North Florida, Jacksonville, Florida, 1988, pp. 233–235.

Daily, John, & Shigemura, Nathan S.: *Fundamentals of Applied Physics for Traffic Accident Investigators* (Vol. 1). Institute of Police Technology and Management, University of North Florida, Jacksonville, Florida. 1997.

Giancoli, Douglas C.: *Physics: Principles With Applications.* California State Polytechnic University, Pomona, Prentice-Hall, Englewood Cliffs, New Jersey, 1980.

Hibbeler, Russell C.: *Mechanics for Engineers: Statics and Dynamics.* MacMillan Publishing Company, New York. 1985.

Howell, Wiley L.: *Derivations Manual for Formulas Used in Traffic Accident Investigation and Reconstruction.* Institute of Police Technology and Management, University of North Florida, Jacksonville, Florida. 1994.

Metcalf, H. Clark, Williams, John E., & Dull, Charles E.: *Modern Physics.* Holt, Rinehart and Winston, Inc., New York, 1964.

Rivers, R. W.: *Speed Analysis for Traffic Accident Investigation* (2nd ed.). Institute of Police Technology and Management, University of North Florida, Jacksonville, Florida, 1997.

Rivers, R. W.: *Technical Traffic Accident Investigators' Handbook* (2nd ed.). Charles C Thomas, Publisher, Springfield, Illinois, 1997.

Smith, Alpheus W., & Cooper, John N.: *Elements of Physics* (9th ed.). McGraw-Hill Book Company, New York, (undated).
Williams, Dudley, & Spangler, John: *Physics for Science and Engineering*. D. Van Nostrand Company, New York, 1981.

# INDEX



*Note:* Terms used in this index are intended to serve as prompters for locating items of interest in the manual, using a paragraph numbering system prepared so as to provide for easy and direct access to an item's precise location. In the references quoted, the number preceding the decimal point designates the chapter number and the following numbers designate the paragraph sequence in the chapter. For example, in the case of reference number *1.003,* the *1* designates Chapter *1* and *.003* designates the third paragraph of that chapter. Similarly, in the case of the number *4.035,* reference is made to Chapter *4* and paragraph *35* thereof. In the case of figures and formulae, *Figure 1–03* (or formula), for example, indicates the *third figure* (or formula) in *Chapter 1.*

## G